MASTERS OF TIME

D1468554

MASTERS
OF
TIME

Cosmology
at the
End of
Innocence

JOHN BOSLOUGH

Illustrations by Wendy W. Cortesi

A William Patrick Book

Addison-Wesley Publishing Company

Reading, Massachusetts Menlo Park, California New York
Don Mills, Ontario Wokingham, England Amsterdam Bonn
Sydney Singapore Tokyo Madrid San Juan
Paris Seoul Milan Mexico City Taipei

Many of the designations used by manufacturers and sellers to distinguish their products are claimed as trademarks. Where those designations appear in this book, and Addison-Wesley was aware of a trademark claim, the designations have been printed in initial capital letters or all capital letters.

Library of Congress Cataloging-in-Publication Data

Boslough, John.
    Masters of time : cosmology at the end of
innocence / John Boslough; illustrations by Wendy W. Cortesi.
      p. cm.
    "A William Patrick book."
    Includes bibliographical references and index.
    ISBN 0-201-57791-7
    ISBN 0-201-62237-8 (pbk.)
    1. Cosmology. 2. Astronomy—Philosophy. 3. Religion and
science—1946–   I. Title.
    QB981.B7269 1992
    523.1—dc20
                                      91-39017
                                          CIP

Copyright © 1992 by John Boslough

All rights reserved. No part of this publication may be reproduced, stored in a retrieval system, or transmitted, in any form or by any means, electronic, mechanical, photocopying, recording, or otherwise, without the prior written permission of the publisher. Printed in the United States of America. Published simultaneously in Canada.

Jacket design by Mike Fender
Text design by David Kelley Design
Set in 10-pt New Century Schoolbook by Carol Woolverton

1 2 3 4 5 6 7 8 9-MA-959493
First printing, April 1992
First paperback printing, July 1993

FOR SUSAN

▲

*It is the customary fate of new truths to
begin as heresies and to end as superstitions.*
—THOMAS HENRY HUXLEY

The eyes are open, but the sense is shut.
—WILLIAM SHAKESPEARE (*King Lear*)

▼

CONTENTS

Once
upon a time, almost everybody believed
that the universe was created several billion years ago
in a gigantic explosion known as the big bang. The two decades from
1965 until the mid-1980s were the heyday of the theory's popularity.
During those years there was very little dissension among either theoretical scientists or the general public from the view that the big bang
actually had taken place.

The big bang was simply a fait accompli. Both the popular and scientific press gave it uncritical acclaim. Newspaper and magazine articles
and television documentaries glorified science's version of Genesis. A
number of popular science books, including one of mine, simply took the
scientists at their word, then went on to report in a highly romanticized
style about the laudatory efforts of the theorists hot on the track of the
big bang and the universe's subsequent evolution.

This is not another one of those books. During the 1980s I traveled
all over the world while preparing a series of science articles for *National
Geographic* magazine on topics such as particle physics, gravity and
time. I spoke with many of the world's most highly regarded theoretical
physicists and visited most of the big science laboratories and observatories throughout the United States, Europe, Japan, Australia, China and
the Soviet Union. As I traveled from country to country, it became apparent in about 1985 or 1986 that something was wrong. A few years earlier
theorists had routinely expressed great confidence in their ideas about
the origin of the cosmos and, closely related, the emerging "theory of everything" that would soon explain all the forces and matter throughout
the universe. Stephen Hawking was certain that finalization of these
theories was just around the corner. So was Murray Gell-Mann, the
Nobel laureate whose work in the 1960s had led to the standard model
of particle physics.

By the second half of the decade nobody, including Hawking, who
was vacillating on several key issues, was sure about anything. Even
more telling was that an increasing number of wildly speculative, even
outlandish, theoretical ideas were being put forth: time travel through
quantum wormholes, time reversal, exploding galaxies, invisible cosmic
strings and knots, parallel and multiple universes, baby universes, undetected particles such as axions and gravitons, multidimensional space,
vibrating superstrings, unseen dark matter and on and on.

Where is all this leading us? Do these ideas paint a picture of a bold
new universe, or just a more chaotic view of the one we already know?
None of the new suppositions has the slightest possibility of being put to

the test of experimentation or direct observation. Yet, strangely, many of the best scientists have proven purblind to these basic tenets of scientific inquiry. What's more, these scientists have succeeded in convincing a willing and eager public that the final grand secret of the universe is all but at hand.

Viewed in perspective, the glowing language and religious metaphor cosmologists used in describing their enthusiasm for the 1992 discovery of tiny variations in the cosmic background radiation actually may have been an inverse measure of the desperation they have been feeling for the past six or seven years. Most astrophysicists maintained that the discovery confirmed the big bang beyond the shadow of a doubt, solving the mystery of the universe's origin and structure once and for all, and all but revealing the very mind of the Creator Himself.

Really? Are we humans any closer now to laying bare the secret of the universe's creation than we were before the discovery? Does such a secret actually exist? And if so, is The Answer all but at hand, as cosmologists suggest?

In fact, we are further from the revelation of such a secret, assuming it actually exists, than at any time in recent history. New astronomical observations have contradicted almost every theory purporting to explain how galaxies formed, the crucial missing link in the history of the universe. At the same time experiments in big atom-smashing accelerators and elsewhere have failed repeatedly to turn up evidence for the unification of forces that is required for the great theory of everything.

This book is the story of what happened along the road to the big theory and how it may be a mere will-o'-the-wisp rather than a final destination. The book also seeks to raise a number of questions about the biggest sacred cow of all, the big bang. What are the leaps of faith made by big bang theorists? Will the big bang really prove to be the ultimate theory about the origin and structure of the universe? Is the big bang fact or convenient fiction?

Some of these ideas may not find a ready audience among cosmologists; many, in fact, will not like what they read here. Other readers may feel the material is too complicated for general understanding or that it is too superficial. I have tried to strike a middle ground in order to make the story accessible to thoughtful members of the general public. Science, after all, is still a process and not a monolith.

Preparation and research for this book, which was written during 1991 in McLean, Virginia and Aspen, Colorado, consumed the better part of the last decade. Many, many people were exceedingly gracious with their time and ideas, some of whom put in appearances in the book and others who do not. I cannot begin to thank them all enough, but I will try.

I appreciate especially the time given me by physicists whose daily calendars are filled to overflowing with the affairs of people as well as the affairs of the cosmos: Eric Adelberger, Bruno Bertotti, Nicolo Beverini, Romano Bizzarri, Roger Blandford, Tom Bowles, Paul Boynton, Vladimir Braginski, Blas Cabrera, Sidney Coleman, Alvaro De Rújula, Mark Dragovan, Ronald Drever, Donald Eckhardt, Bob Edge, Francis Everitt, the late William Fairbank and his colleagues at Stanford, James Faller, the late Richard Feynman, Ephraim Fischbach, Harold Furth, Murray Gell-Mann, Sheldon Glashow, Terry Goldman, Alan Guth, Stephen W. Hawking, Drasko Jovanovic, Leon Lederman, and Charles W. Misner.

Also: Riley Newman, Michael Martin Nieto, Jeremiah Ostriker, Jim Peebles, Jeffrey B. Peterson, Pio Picchi, Roger Penrose, Burton Richter, Carlo Rubbia, Keith Runcorn, Donald Schneider, Frank Stacey and his colleagues at the University of Queensland, Samuel Ting, Jim Thomas, Edwin Turner, Gerald Wasserburg, John Archibald Wheeler, Madame Ye Shu Hua and her colleagues at the Shanghai Observatory, George Zweig, and Mark Zumberge.

The work of noncosmologists also helps sort out the mysteries of the universe. Among those who directed me along fruitful paths were William Andrewes, former curator of the Time Museum in Rockford, Illinois, and Seth Atwood, the museum's founder; cosmonaut-physician Oleg Atkov of the Academy of Medical Sciences' Department of Clinical Cardiology in Moscow; General Vladimir Dzhanibekov, cosmonaut and hero of the Soviet Union; Dr. Oleg Gazenko, director of the Institute of Biomedical Problems in Moscow; David W. Allen and his colleagues at the Time and Frequency Divison of the National Institute of Standards and Technology; Charles Ehret of the Chronobiology Division of Argonne National Laboratory; Dr. Ralph Pelligra, Dr. Harold Sandler and Dr. Emily Holton of the Medical Research Division at NASA's Ames Research Facility; Bob Williams of NASA's Zero-Gravity Program and Dr. Samuel Pool of NASA's Weightless Training Center at the Johnson Space Center; Professor Stuart Malin of the Old Royal Observatory in Greenwich, England; and Professor Rupert Hall formerly of the Imperial College in London.

Others were Dr. John B. West of the University of California at San Diego; Silvio A. Bedini, specialist on time at the Smithsonian Institution; Harald Bungarten, Roger Antoine and Vince Hatton of the European Organization for Particle Physics (CERN) in Geneva, Swizerland; the late, great Margaret Pearson of Fermilab; Janet Dudley, Graham Appleby and Peter Standen of the Royal Observatory at Herstmonceaux, England; Richard Hills of the Greater Manchester Museum of Science and Industry in England; Farouk El-Baz of Boston University; Michael Mahoney, professor of the history of science at Princeton; and Professor Thomas Kuhn of the Massachusetts Institute of Technology.

Also: Ronald Harper of the Brain Research Institute at the University of California at Los Angeles; Dr. Peter Hammond of the Anthropology Department at UCLA; C. Randall Morrison, an archeologist with the Bureau of Indian Affairs in Phoenix; Dr. Bernard Guinot, former director of the Bureau Internationale de l'Heure in Paris; Dr. Suzanne DeBarbat of the Paris Observatory; P. David Seaman, an anthropologist at Northern Arizona University; Hiroshi Tsukahara, Koji Okura and Kenji Fujita of the Seiko Group in Tokyo and Suwa, Japan; Katsuhiro Sasaki of the Department of Physical Sciences at the National Science Museum in Tokyo; Dr. Wu Guei-chen and his colleagues at the Shannxi Observatory near Xian, in the Peoples Republic of China, and Dr. Miao Young-Rui and Dr. Lo Ding-Jaing of the Beijing Observatory; Janine Perret of the Ebel watchmaking group in La Chaux-de-Fonds, Switzerland; and Gernot M. R. Winkler and his colleagues at the U.S. Naval Observatory in Washington, D.C.

Wendy Cortesi brought thorough research and thoughtful questions to the manuscript-in-the-making while Rebecca Withers produced quick answers when they were needed. Providing me with the means and time to visit some of the world's most interesting places and meet some of its most enlightened inhabitants while beginning research on this book were my editors and others at the National Geographic Society, among them Bill Garrett, former editor of *National Geographic* magazine, Tom Canby, Charles McCarry, Joe Judge, Marie Barnes, and Lilian Davidson.

Others provided support and inspiration or a combination of the two. I can't begin to thank them enough: Dr. James Boslough, Katherine Gibson Boslough, Susan Boslough, Jill Campana, Kenneth Garrett, Lucie Morton Garrett, Dick and Nancy Johnson, Jim Sugar, May Raehn, Ray Raehn, Trent J. Bertrand, John Brockman and his associates in New York City, Jan Adkins, Carlota Shea, Jayne Wise, Dan Moldea, Mimi Wolford, Hensley and James Peterson, Darryl G. (Taro) Kaneko, and the Lhasas Boomer and Sasha, who were with me every step of the way during the final hectic stages of writing, reading and revision.

I would like also to sound a note of sincere appreciation to my editor, William Patrick, who was adept at finding the middle ground between the stick and the carrot and unusually sensitive when it came to deciding which to wield at a particular moment and how strongly.

PART I

THROUGH A TELESCOPE DARKLY

*I want to know
God's thoughts.
The rest are details.*
—ALBERT EINSTEIN

God is in the details.
—MIES VAN DER ROHE

▼

CHAPTER **ONE**

CRISIS
IN THE
COSMOS

*So the first thing, if I may suggest it, is that
you completely stop inwardly. And when you do stop
inwardly, psychologically, your mind becomes very
peaceful, very clear. Then you can really look at
this question of time.*
—KRISHNAMURTI

On
a blustery winter day in Chicago
six hundred physicists and members of the media
packed a lecture hall near Lake Michigan. It was December 1986. The
atmosphere inside was heavy with anticipation. Just outside the audito-
rium the most celebrated theoretical physicist of his generation, Stephen
W. Hawking, sat slumped in his motorized wheelchair cracking a last-
minute joke with one of his graduate assistants.

As he was introduced to thunderous ovation, Hawking wheeled
onto the proscenium. He looked uneasy and unusually wan in the glare
of the spotlight. The audience fell silent and stared expectantly at him.
With an all but imperceptible flicker of a thin finger, he pressed a button.

"Today I want to talk about the direction of time," he said, speaking
in a metallic, clipped monotone through a specially fitted voice synthesizer. "What is the difference between the past and the future? Why
do we remember the past, but not the future? And how is this connected
with cosmology?"

The specially fitted computer was designed for a person lacking the
strength and control to type, someone who can make only one movement.
Hawking had suffered from a degenerative neuromuscular condition for
nearly three decades. When he wrote his lecture, the cursor on the screen
flicked among letters of the alphabet, stopping at one when he squeezed
his switch. This called up a screen full of pre-programmed words that
began with the letter he selected.

There were about 2,600 words in the computer. They combined the
esoteric language of theoretical astrophysics with the mundane vocabu-

3

lary a person needed to get around in the world: theory, thermal, thanks, topology, tea, thermodynamics, Thursday; quirk, quark, quiet. The cursor scrolled down the list until Hawking saw what he wanted; the word was added to a sentence at the bottom, ready to be pronounced through the speaker at the back of his wheelchair or, for precise note taking, displayed on a separate computer monitor.

Hawking was in Chicago to attend an international physics conference. He was just then attaining worldwide celebrity. His photograph had appeared in newspapers and magazines, and television crews followed him everywhere he went as if he were visiting royalty or a rock star. As he zipped down the street at the University of Chicago in his wheelchair, onlookers saluted him with clenched fists and shouted, "Right on." He whirred around the dance floor of a nightclub as delirious dancers made way for him.

Hawking was Lucasian Professor of Mathematics at England's University of Cambridge. The chair had once been held by Isaac Newton, and the current holder liked to observe wryly that it was obvious that the chair he used was not the same one Newton had occupied. Hawking had been responsible for several major theoretical breakthroughs in the 1960s and 1970s that had helped shape the way physicists looked at the universe.

Then in 1985 he had startled his fellow theoreticians by declaring that the direction of time would reverse under certain special conditions. This would occur, he said, when the expanding universe of today stopped growing and began contracting. Has any statement from a major figure more clearly embodied the supreme confidence—if not hubris—of modern science? The reversal of time? How much further out along the border between theoretical possibility and fantasy could cosmology go?

In the Chicago lecture hall, Hawking's computerized voice began droning out his pre-programmed talk in an accent sounding vaguely like that of a Hungarian count who had studied English in Canada. Many in the audience may have guessed that he once again was about to defend his highly speculative time-reversal proposition. A few of his closest colleagues knew better.

The End of Innocence

Hawking was about to announce a major change of mind that would help to signal the end of humanity's age of innocence about the cosmos. Hawking had personified that innocence. In a conversation in 1981, he had confidently assured me that by the end of the century he and other cosmologists would have created a single theoretical statement that would describe not only all the physical laws of the universe, but also its initial conditions.

"My goal is nothing less than a complete understanding of the universe, why it is as it is and why it exists at all," he said. He was himself amused by the wonderful audacity and presumptuousness of the remark. When I saw him again a year later, the first thing he said to me was, "My goal is nothing less than a complete understanding of the universe." He said this with great mock seriousness, but he didn't disavow it; in fact, he had been passing out photocopies of the magazine article in which the original statement of his goal had appeared.[1]

The only remaining problems for physicists, Hawking and most of his colleagues believed in the early 1980s, were purely theoretical. The theorist's job was to provide daring new theories that would make predictions that could then be tested by observational astronomers or physicists working in laboratories. Theorists expected that more and better observations would continue to clarify the picture of the cosmos.

What nobody realized in those days was that just the opposite would be the case. New discoveries in the decade ahead would lead not to a clear statement of a final secret of the universe, but instead to a confused and poorly composed picture of a cosmos more idiosyncratic than anyone had imagined. By 1990 nobody, except for a few diehards who persisted in believing that revelation was just around the next corner, was talking about a final secret anymore.

The principal cosmological model, the big bang theory of the origin and evolution of the universe, was vexed with major problems throughout the late 1970s and 1980s. More and more, its theoretical underpinnings were inconsistent with new astronomical observations. For example, one major finding in the mid-1980s was that galaxies were spread across enormous bubblelike structures hundreds of millions of light years across. This suggested that the universe was lumpy across vast regions of space.

Where did the galaxies and great clusters of galaxies come from? A number of theories have sprung up to explain the distribution of galaxies, but none of them has provided a satisfactory scenario. Scientific discoveries have dashed the hopes of theorists over and over. Surprisingly, these theorists have not proved to be very resilient. Most seem reluctant to abandon their ideas even when confronted by new observations in direct opposition to their theories.

They have preferred to modify a pet theory with ad hoc suppositions and tacked-on assumptions or to build into their computer models of the universe adjustable parameters that can easily accommodate inconvenient facts rather than simply to discard a theory that no longer works. This has led to a dangerous trend in which many theorists have accepted new hypotheses long before they can be tested by experiment or, worse, new hypotheses that have no hope of ever being tested.

Another troubling trend is that physicists have resorted to concocting increasingly complicated mathematical hypotheses to explain the universe. They justify these excruciatingly difficult equations on the basis of their "elegance" or "inner truth and harmony," all reminiscent of a Platonic attitude that the universe can be fathomed by pure thought alone.

Among these concepts are the inflationary universe theory, string theory, wormholes, dark matter and shadow matter, parallel universes and rolled-up dimensions. Except for the inflationary universe theory, which hypothesizes a short burst of exponentially rapid expansion when the universe was less than a second old and which did make one prediction, later disconfirmed, the unifying theme of all these new ideas is that none has yielded a single testable prediction. In the case of one of Stephen Hawking's latest conjectures, the so-called no-boundary theory, which attempts to skirt altogether the notion that the universe must have had an absolute beginning, the idea is so abstract that it will never be tested during the lifetime of anybody now alive.

In the late 1960s particle physicists, who study the fundamental entities of matter and the forces that draw and repel these entities, joined forces with cosmologists. Initially there were good results, notably an explanation for the preponderance of the light elements, helium and hydrogen, throughout the universe. The so-called standard model used by particle physicists has been moderately successful in describing the basic units of matter and the four fundamental forces of nature—gravity, electromagnetism, and two powerful forces in the atomic nucleus. Yet the theory requires a score of adjustable parameters.

Extensions of this standard model that attempt to unify three of the forces, such as the grand unified theories, or GUTs, are far out on the limb of untestability. The one prediction of the various grand unified models that did seem to be testable—that is, that every proton in the universe eventually would decay—has led to a number of ongoing experiments that have utterly failed to turn up a single shard of evidence that the models were correct.

The Cosmic Priesthood

Beginning shortly after the Renaissance, scientists adhered to an approach in which they first looked at the world around them, then developed hypotheses about how things seemed to fit together. The hypotheses were constructed in such a way that they made specific predictions that could then be tested directly to determine the truth of each hypothesis. In truth, scientific development is not always this simple, yet cosmologists appear to have abandoned this so-called scientific method as if it were useless and annoying.

Cosmology, of course, is the least practical of the sciences, and the most sublime. A scientific endeavor that held out as its goal a complete portrait of the history and structure of the universe would have to tolerate unusual procedural methods, cosmologists seemed to be saying.

Moreover, cosmologists had become members of an exclusive community that was the perfect priesthood for a secular age. They, not religious leaders, were the ones who would now reveal the secret of the universe bit by precious bit, not in the guise of spiritual epiphany but in the form of equations obscure to all but the anointed.

By holding them in awe, the rest of us aided, abetted and enabled the high priests of cosmology. In *Lonely Hearts of the Cosmos,* Dennis Overbye, a self-described cosmological camp follower, encounters Stephen Hawking.

"In the end what I wanted to know from Hawking was what I have always wanted to know from Hawking," Overbye writes. "Where we go when we die."

Hawking, of course, could only decline to answer, but Overbye persisted. "If we couldn't see God, could we at least know God was there, even if sulking in a black hole or at the end of time? What I wanted from Hawking was some touch of the miraculous." [2]

That was what the public at large seemed to want from the cosmologists, a touch of the miraculous. In our hearts we were all cosmologists with ideas about the origin and shape and size of the universe. But we had yielded our right to wonder about the cosmos to a handful of mathematical prodigies.

By the beginning of the 1990s, though, instead of a miraculous revelation, cosmology was encountering its first fullfledged crisis. A multitude of astronomical discoveries had combined with one experimental failure after another to overwhelm the best-laid theories. In jeopardy were the various grand unified theories (GUTs) as well as the theories of everything (TOE), which purported to describe the events of the very early universe.

All this theoretical uncertainty has begun catching up with the big bang theory itself, the leading model of cosmic genesis and evolution. Proponents of the theory have begun rushing to its defense, a sure sign of trouble afoot.

"We have trouble predicting tornadoes, but that doesn't mean that we in any way doubt that the earth is round," said David N. Schramm, a cosmologist at the Fermi National Accelerator Laboratory near Chicago and one of the big bang's chief paladins. "Similarly we have trouble making galaxies, but we don't doubt that there was a big bang." [3]

The analogy, as Schramm no doubt realized, was deeply flawed: We know the earth is round; moreover, tornadoes relate only meteorologically to that fact. In a 1991 article called "Big Bang Bashing," which condemned the practice, the popular science magazine *Discover* defended the big bang theoretical model. But the magazine did concede that the theory "will last only until someone makes an observation that convincingly rules it out, or until someone creates a model that both explains the facts better and also makes unique predictions that are later borne out." [4]

We seem to forget that when the big bang theory was created during the middle decades of the twentieth century, it was considered a good workmanlike model that seemed to explain a few astronomical observations. It was not thought of as the final grand secret of the universe. But the model's press, including an endorsement by the pope, was extraordinary. The big bang became such an important part of popular consciousness that it was easy to forget that it was nothing more than a theoretical model, and, as the years have passed, an increasingly endangered one at that.

A shift in thinking has already begun, with several new ideas cropping up. It is only a matter of time until the big bang collapses. It is doubtful that anybody looking back from the perspective of the next millennium will regard the big bang as anything more than a quaint theoretical backwater, lost in the tides of history.

CHAPTER TWO

THE
MOVABLE
SKY

Now this is the Law of the Jungle—
as old and as true as the sky.
—RUDYARD KIPLING

Cosmological
speculation from Plato to Hawking
went back to the ancient sky, the single visible constant
in the life of every person inhabiting the earth. No culture could ignore
the heavens. For ancient Norse sailors the sky was "the path of the
ghost," for the Navajo "the backbone of the night." In the ancient civilizations of China, of the Maya and the Greeks, astronomy was the first
science, along with its odd little cousin, astrology, which seemed to ex-
ploit the mind's preoccupation with finding connections between unrelated events.

One of the most unusual pieces of evidence for humanity's eternal
fascination with the sky stands in northern Kenya. Some archaeoastro-
nomers, scientists who combine a study of astronomy with archaeology,
believe that an ancient array of nineteen stones at a place called
Namoratung'a is one of the first observatories in Africa. The people who
built it may have come from eastern Asia about three thousand years ago
in the great migration of the Galla people, the first settlers of northern
Africa.

The stones of Namoratung'a marked the horizon positions for the
rising moon and star groups from which its builders formulated early
calendars. Like other prehistoric observatories in Europe and Meso-
america, it was a simple system for decoding a complex array of patterns
in the sky above.[1]

The Namoratung'a observatory was apparently built to predict the
conjunction of the new moon with a few significant patterns, in this case
constellations. However, it only worked for a century or so after it was
built in about 300 B.C. because of the phenomenon known as the preces-

9

Namoratung'a   Early astronomers may have used these stones to make predictions about the alignment of stars and the moon.

sion of the equinoxes. Caused by the gravitational attraction of the sun and moon on the Earth at the equator, the stars appear to move slowly westward in a cycle that takes 25,800 years to complete.

After about a hundred years, the predictions of the stones were no longer fulfilled. The skywatchers who used the observatory had to make what was undoubtedly an unhappy decision to abandon it. The story was the same throughout history. No matter the time or the place, astronomers frequently encountered unexpected new patterns in the night sky, forcing them to give up their cherished models.

In the fourth century B.C. Aristotle built a comprehensive model of the cosmos. Like virtually every ancient model, it was based on the assumption that the Earth lay at rest in the center, with the sky revolving around it once during the course of a day and night. Aristotle's model was refined and codified in the *Almagest* by Ptolemy, the Greek scientist who worked at Alexandria during the second century A.D.

Under Ptolemy's system, used by astronomers for nearly a millennium and a half to calculate the positions of celestial objects, most heavenly bodies were carried about the Earth on their own rotating circles. That the sun might not be circling around the earth must have simply defied common sense. The first known break from that entirely logical view came in 1508 when the Polish astronomer Copernicus, after seeing that the patterns in the sky had shifted, first dared write the words, "What appears to us as the motions of the Sun arise not from its motion but from the motion of the Earth."

Such thinking was so alien to the spirit of the times that Martin Luther wrote, "The fool will turn the whole science of astronomy upside down." Which is precisely what Copernicus proceeded to do in his codification of a heliocentric solar system in *De Revolutionibus Orbium Coeles-*

tium published in 1543, the year he died and the year some historians think marked the onset of modern science.

Late in the seventeenth century the great English scientist Isaac Newton explained how gravity was responsible for keeping the planets in the orbits that had been described by Copernicus and endorsed by Galileo Galilei, who had been condemned by the Church for his heresy. The next major overhaul in astronomical thought did not occur for nearly four centuries. In the years during and after World War I American astronomer Harlow Shapley measured the distances to the Cepheid variables, pulsating stars fluctuating regularly in brightness in our galaxy, and concluded that they lay at enormous distances from the sun: between 15,000 to 100,000 light years away.*

Before Shapley, the most distant stars had been measured at only about 100 light years. His discovery was a devastating blow to the egocentric view that humanity occupied a central place in the cosmos. Yet matters were only to get worse for a lonely species trying to grasp some kind of understanding about its role in the cosmos.

Using the new 100-inch telescope on Mount Wilson in California, in 1926 Edwin Hubble, a former boxer, Rhodes scholar and law student turned astronomer, showed that the Milky Way did not comprise the entire universe, still another setback to humanity's ego. A few years later Hubble analyzed the light from the most distant star systems and found them receding from our solar system at enormous speeds, the first observational evidence that the entire universe was in motion.

One possible consequence of a modification of Albert Einstein's theory of general relativity that had been undertaken in 1922 by the Russian mathematician Alexander Friedmann was that the universe was expanding. Hubble's discovery further complicated the picture of the universe and led to a controversy that was not resolved for decades. Was the universe, in fact, expanding, as most observational and theoretical astronomers believed, or was it in a steady-state that only appeared to be expanding? If the universe were expanding, then what could it possibly be expanding from?

During the middle years of the century, physicists arrived at an answer. A Belgian physicist-priest named Georges Lemaître first called it the primeval atom. Later the primordial fireball of nearly infinite density from which the universe and its contents supposedly evolved came to be known as the big bang.

---

*The speed of light is believed to be 186,281.7 miles per second in all parts of the universe. During a year, light travels 5,880,000,000,000 miles. In 1838 Friedrich Wilhelm Bessel, an astronomer at the Königsberg observatory in Prussia, first applied the concept of a light year to measuring astronomical distances. He showed that a faint star called Cygni 61 lay at a distance of 11 light years (about 60 trillion miles). The diameter of our Milky Way galaxy is about 100,000 light years.

The Hubble Expansion   Objects farther from Earth appear to recede at a faster velocity. As the universe expands at a uniform rate from position 1 to position 2, galaxy B moves farther from Earth than galaxy A. Thus, the velocity of galaxy B relative to Earth is greater.

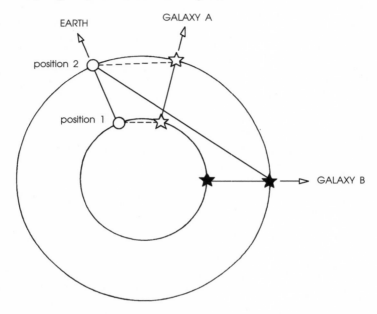

In 1965 scientists at Bell Laboratories in New Jersey accidentally discovered uniform signals from all parts of the sky. For most scientists studying the evolution of the universe, this was all the evidence they needed for what seemed to be relic radiation from the energetic, dense big bang 10 billion to 20 billion years or so ago. Later measurements by equipment carried aloft by high-altitude balloons or into earth orbit by spacecraft confirmed the startling uniformity of this background radiation, which by the 1990s became the source of an increasingly irritating problem for cosmologists. The big bang theory seemed to be confirmed both by the evidence of the receding galaxies and by background radiation. Physicists and the public embraced the theory. It was easy to understand, overpowering in its simplicity and seemingly divine in its union of modern theoretical physics with metaphysics and religion.

The Traveling Galaxy

Ironically, the unraveling of the big bang began shortly after its confirmation. In the 1970s an astronomer named Vera Rubin began turning up evidence that the universe was changing faster, more dramatically and in ways not consistent with the normal Hubble expansion.

The role of iconoclast came naturally enough to Rubin. When she graduated from Vassar in 1948 and wrote away for the catalog and application forms to the highly regarded graduate program in astrophysics at Princeton, her letter was not even answered. Princeton was still twenty-three years away from accepting women into its graduate school.

She attended Cornell instead, where she studied astronomy and worked in the physics department. Cornell was a very exciting place to be in those days because of the other irreverent minds working there: Philip Morrison, Hans Bethe, who later won a Nobel prize for his work on the fusion dynamics of the sun and stars, and the irrepressible Richard Feynman, who was to receive a Nobel for leading the way in the creation of quantum electrodynamics.

At Cornell Rubin became curious about the movement of galaxies. At the time the notion of a rapidly expanding universe was fairly well accepted by most physicists. The exceptions were a cadre of holdouts led by British theorist Fred Hoyle, who believed that the universe existed in a steady-state rather than one of rapid expansion. Still, the notion that the expansion had begun in the big bang was becoming a pervasive one throughout the astronomical community.

Although among the theory's early supporters, Rubin had begun wondering. Do the galaxies move any way other than outward from us in the universe's rapid expansion? Might they rotate in a grand circle around the entire universe the way individual stars circle around a galaxy? Too young and too female to get any observing time at one of the big telescopes, Rubin put together a paper based on data that already had been obtained on 108 galaxies with known rotations. Even by the standards of the day, this was an incredibly small number with which to work.

Hubble and other astronomers already had figured out that light waves from galaxies beyond our own Milky Way were distorted out of shape by the time they reached us, shifted toward the red end of the spectrum. This redshift, which remained controversial, was generally believed to be a measure of the galaxy's speed away from us as well as its distance from us. For her master's thesis, Rubin eliminated the redshift motion from each of the galaxies and analyzed the residual movement and concluded that the galaxies did indeed display large-scale radial velocities in addition to the motion due to expansion.

She entitled the paper "Rotation of the Universe," and presented it as her professional debut at the meeting of the American Astronomical Society. Almost everybody in the astronomy community thought the title was a bit presumptuous for a twenty-two-year-old graduate student.

First, the data were too scanty. And second, the great Kurt Gödel, the world-famous mathematician at the Institute for Advanced Study in Princeton and Einstein's friend, was working on the rotation of the universe. Who was this young female to take on such a project? When the abstract of the paper was printed in the *Astronomical Journal,* its title was changed to the less incendiary "Differential Rotation of the Inner Metagalaxy," but the paper itself was rejected.[2]

After Cornell Rubin moved to Washington, D.C., where her husband, Robert Rubin, a physical chemist, had gotten a new job. She entered the Ph.D. program at Georgetown University and began studying spectroscopy and radio astronomy. This was a homecoming of sorts. Growing up in Washington, she had spent hours looking out her bedroom window, watching the nightly rotation of the heavens around the North Star, and had decided on her life's work—to study the universe.

One day at Georgetown she was contacted by George Gamow, who was interested in her work at Cornell on the rotation of galaxies. Born in Odessa, Russia, in 1904, Gamow had studied at the University of Petrograd (later Leningrad) with the theorist Alexander Friedmann, who had solved Einstein's field equations of general relativity and had mathematically speculated upon the ways in which the universe might be expanding. After receiving his doctorate, Gamow went to Göttingen in Germany where he performed so brilliantly in particle physics that he attracted the attention of Niels Bohr, the legendary theorist who had helped create quantum mechanics, the branch of physics devoted to the fundamental particles of matter and the forces that control them.

Bohr invited Gamow to spend some time at Bohr's new physics institute in Copenhagen. In 1931 Gamow returned to the Soviet Union to renew his passport. Prevented from leaving by the repressive Stalinist regime, Gamow nonetheless managed to persuade the authorities to let him attend with his wife a scientific meeting in Brussels in 1933. He never returned.

Although not especially adept at everyday matters such as spelling or even mathematics, Gamow possessed an extraordinary genius for broaching new ideas and asking penetrating question. In 1928 while in Germany he had formulated an early theory about radioactive decay and was one of the first physicists to address the problem of how stars evolved. He even proposed a significant theory on how genetic information was organized in living cells. At the time Rubin met him he was a professor at George Washington University in Washington and already well known for his popularization of physics in books such as *Mr. Tompkins in Wonderland,* and *One, Two, Three . . . Infinity.*[3]

Gamow was interested in how galaxies were spatially distributed in the universe. At his first meeting with Rubin he fell asleep several times, then woke up and asked questions she considered stupid. Nonetheless, he expressed an overall understanding, which she considered profound, of the problem of galactic distribution. She decided to go ahead with the problem for her doctoral thesis: Were the galaxies arranged randomly? Or was there a pattern that had yet to be discerned?

Because of the limitations of computer power in the 1950s, the problem was daunting. Rubin spent month after month struggling through long and messy calculations on a desktop calculator. At last she concluded there was a very definite clumping of galaxies in the universe with enormous voids of empty space in between. When her thesis was completed in 1954, almost nobody noticed. But the paper set the stage.[4]

During the 1960s, while establishing her credentials as a fine young astronomer at the recently opened Kitt Peak Observatory near Tucson, Arizona, Rubin became the first woman given permission to use the instruments at the Palomar Observatory near San Diego. In 1965 she joined the Carnegie Institution in Washington, where she began a collaboration with physicist and astronomical-instrument designer W. Kent Ford.

Led by Maarten Schmidt, astronomers at Palomar in 1960 had undertaken a search for the most distant objects in the universe. They had discovered quasars, billions of light years away and by far the most distant objects in the universe. They were also flying away at a nearly unbelievable rate, 90 percent of the speed of light.

Like almost everybody else, Rubin and Ford started working on quasars, locating and measuring the redshifts of the newly discovered objects. It was a wild, competitive time. Dozens of astronomers were fast on the tracks of the new objects. The disruptive telephone calls and the uncertainties of working with a hot, new topic unsettled Rubin and Ford. So they went back to the subject that originally had aroused Rubin's curiosity, the motion of galaxies independent of the outward expansion of the universe.

Ever since Galileo improved upon the first telescope and turned it heavenward in the early seventeenth century, skywatchers had always tried to look farther and farther out into the universe. The greatest glory was reaped in the most distant fields. This was where Shapley and Hubble had looked after World War I. It was where Maarten Schmidt had found the first quasar. During the late 1960s and early 1970s most astronomers paid almost no attention to nearby galaxies.

Only a few astronomers had noticed what looked like it could be a poor correlation between the redshifts of these galaxies and their observed distances.* Nobody was worried. Almost everybody simply be-

---

*Astronomers use the principle of parallax to calculate the distance to planets and nearby stars. If the object whose distance is to be measured is observed against a background of more remote objects, its position will seem to move as does the position of the observer. Knowing the distance between two observation points—say, between one side of the Earth and the other or, in the case of distant stars, between one side of the Earth's orbit around the sun and the other side six months later—astronomers can measure the respective angles and easily calculate the distance using trigonometry.

lieved that these discrepancies were the results of technical failures in their measuring techniques and eventually would be corrected. It was all but forgotten that Vera Rubin had shown two decades earlier in her disparaged master's thesis that galaxies could move independently of the universe as a whole, which could explain inconsistencies between observed redshifts and distances.

Rubin and Ford drew back from the momentary infatuation with quasars and other popular, exotic new objects such as black holes, countering the three-century-old passion for looking ever farther out into the universe. They decided to refocus their telescope on the sky closer to home. The results were astonishing.

Attraction of the Dark

Ever since Hubble's discovery that the universe was expanding, astronomers had realized that galaxies in the so-called local group roamed about space with what appeared to be a mind of their own. These galaxies seemed to be moving independently of the general expansion, and astronomers started calling this phenomenon peculiar motion. Containing just under thirty galaxies, the local group encompassed the Triangulum Spiral; the two Magellanic Clouds visible in the southern sky; about two dozen dwarf galaxies; Andromeda, the largest by far; and our own Milky Way, a poor second in size. There are two other possible members, both huge, but they are obscured by dust and the plane of the Milky Way, which slices through space between them and Earth's orbit; observers cannot see them directly.

Astronomers already realized that a few of the galaxies within the local group moved toward rather than away from us due to the influence of the Milky Way's great gravitational pull. The peculiar motions of these and other galaxies were considered too small to be of any great importance. If these motions were large enough to mean anything, astronomers reasoned, then the light of these galaxies would display a shift toward the blue end of the spectrum, meaning they were moving toward us rather than away. This was clearly not the case.

During the early 1970s Rubin and Ford measured the peculiar velocities of the wandering galaxies. One day in 1975 they made a remarkable find. Not only did nearby galaxies have a surprising amount of extra motion, but the Milky Way itself had a velocity of about 500 kilometers

---

In the case of stars or galaxies that are more than a few hundred light years away, spectroscopic methods reveal the real luminosity of a star, which is then compared to its observed brightness to obtain the star's distance. A Cepheid variable is a kind of pulsating star whose period is related to its real luminosity, so its distance can easily be calculated from its apparent brightness. Cepheids in nearby galaxies are believed to give highly accurate measurements of their distances.

per second that had absolutely nothing to do with the expansion of the universe. The motion, instead, was in relationship to a group of distant galaxies, and was decidedly more _peculiar_ than most astronomers had ever imagined possible.[5]

Critics were immediately wary of the so-called Rubin-Ford effect. In those days the overwhelming consensus was that the expansion of the universe was going on smoothly, quietly and without undue disruption. It would upset the established theories to find any large-scale deviations in the velocities of individual galaxies. Rubin and Ford were criticized roundly at conferences. Better-known astronomers warned them to drop their research.

The critics acknowledged that the Milky Way might seem to be wandering through the cosmos in the wrong direction. But actually, they said, this was an illusion of movement; it undoubtedly was the result of incorrect measurements Rubin and Ford had taken of the distances between galaxies. A few senior colleagues advised her to desist if she knew what was good for her career. Discouraged, Rubin quit.

"We interpreted what we observed as a large motion of our galaxy," she said. "I guess that was not easy for some people to believe."[6]

She turned her attention back to her old work in galactic dynamics. Margaret and Geoffrey Burbridge, a wife-and-husband team at the University of California at San Diego, had started looking at the centers of galaxies. So Rubin and Ford decided to complement the Burbridges' work by focusing on the spiral arms. She reasoned that the way a galaxy was spinning would surely go a long way toward explaining its basic structure.

Some galaxies had tightly wrapped arms that, like a spinning skater, might mean the galaxy was rotating rapidly. Others had arms that, opened wide, spread thousands of light years across the sky. Rubin wondered why.

Kent Ford recently had designed a sensitive, new spectrographic image tube that was perfect for the job. Fixed to a telescope, the tube had a slit that could be lined up with the disk of a spiral galaxy in order to measure whether the light from its arms shifted toward the red or blue end of the spectrum. Most galaxies appeared to have huge, bright bulges at their centers consisting of dense concentrations of stars. The spiral arms were usually mere wisps of luminosity, emaciated by comparison with the chunky galactic torso.

Astronomers assumed that the standard pattern meant that the largest concentration of the galaxy's mass was also packed around its core. Thus, a galaxy should behave exactly like a gigantic solar system: the great gravitational attraction of the central mass keeping the outer objects in place and setting their velocity. According to Newton's inverse-square law, the close-in objects should literally gallop around the center while the more distant objects would rotate at a more stately pace. This

Spectrograph and Spiral Galaxy   Light from the side of the galaxy rotating toward Earth is shifted to the blue end of the spectrum, while light from the side rotating away from Earth is shifted to the red end.

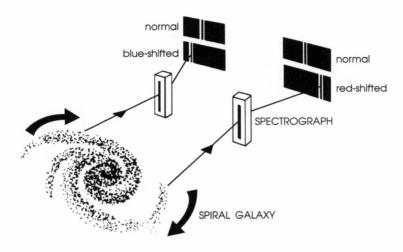

was true of our solar system. There was no reason to expect that it would be any different for any galaxy considered as a single entity.*

Rubin and Ford first aimed their new instrument at Andromeda. The nearest galaxy, by good fortune, happened to have its spiral disk perfectly inclined toward Earth to match the slit in their spectrograph. The analysis of its spiral arms took long hours with noncomputerized instruments. When they were done with their analysis, they were astounded by what they found. The stars and gas clouds on the outskirts of Andromeda's spiral arms were rotating every bit as fast as those at the galaxy's center.

This could be true only for Andromeda, they felt sure. It must be a very strange galaxy, indeed. But as they began examining other spiral galaxies, they discovered the same thing over and over: The outer stars and gases moved every bit as fast as those near the core. It looked like a physical impossibility.[7]

Something was very wrong. What could possibly be holding the galaxies together? If Newton's inverse-square law were correct, the outer stars and gases should simply have twirled off into space eons ago, orphans on a bleak and endless journey out into the void. That had not

---

*Isaac Newton's law of gravitation stated in its most simple form that the mutual attractive force between two bodies is proportional to the product of their masses divided by the square of the distance between them.

happened. Rubin realized that gravitation was the only force strong enough to keep the galaxies intact across their vast breadth. And where there was gravity, she knew, there had to be mass in the form of some kind of matter.

But what was it? Whatever it was was invisible to optical telescopes. Rubin and Ford analyzed over two hundred galaxies in the late 1970s and early 1980s. The results left little room for doubt: There was extra matter, unknown material of unknown origin and unknown dimension, in virtually every galaxy they had examined. But more mysterious still, their analysis revealed, this hidden galactic stuff was at least ten times as massive as the luminous stars and dust they could see with their telescope.[8]

Somehow, over 90 percent of the matter in the universe had not been accounted for.

CHAPTER THREE

U<small>P</small>
A<small>GAINST</small>
<small>THE</small> W<small>ALL</small>

*The fault, dear Brutus, is not in
our stars, but in ourselves.*
—W<small>ILLIAM</small> S<small>HAKESPEARE</small>

V<small>era</small>
Rubin and Kent Ford had turned
up the first piece of evidence that most of the matter
in the universe was not the bright, shining stuff of the stars. They had
already made news with their discovery—which the collective judgment
of the astrophysics community had summarily dismissed—that our
home galaxy was traveling through space at an enormous velocity, completely unrelated to the general expansion of the universe. If Rubin and
Ford were right and their models of cosmic growth were correct, the
prodigal Milky Way was, in essence, racing around a corner when it
should have been running down the street.

By the early 1980s most people in the astrophysics community were
hardly alarmed by these minor, though vexing, inconsistencies. They
were seen as merely the splatter of rain on the surface of a deep, tranquil
sea. Yet within just a few years Rubin and Ford were to be vindicated—
more or less. And by the end of the decade most astronomers were gazing
uneasily out into the cosmos at the strange, shifting patterns that the
two iconoclasts from the Carnegie Institution had first revealed.

Patterns in the Sky

Astronomers have always searched the night sky for meaningful
patterns. But as many skywatchers had painfully learned, a pattern could
be a dangerous thing. It could obscure the truth rather than reveal it.

Long before civilizations emerged, our earliest ancestors looked
heavenward and saw constellations. Like anybody else, they must have
found it impossible to look at the night sky without seeing patterns. Try

looking at the seven stars in the northern sky that you have known all your life without seeing the Big Dipper. In the southern sky, you can't help but see the Southern Cross formed by four unusually bright stars.

Most of the patterns are visual groupings created by line-of-sight observations such as those taken by the observers at Namoratung'a. Only instruments more discerning than the human eye could establish that the stars in a particular constellation were not necessarily related. The star at the end of the Big Dipper's handle, Alkaid, is more than twice the distance from Earth as the second star, a binary known as Mizar.

About 3000 B.C. Babylonian kings ordered the drawing of maps of their realm in order to record the tenure of land; one map still survives on a clay tablet. Attempts to chart the sky roughly paralleled the progress of cartography. Sumerian astrologer-priests constructed the earliest known maps of the heavens at about the same time. Looking up, they imagined that they saw a zodiac (literally, a circle of animals), a region of the night sky through which the sun appeared to pass during the course of a year. Using a system of numerology based on six, they divided the zodiac into twelve zones corresponding approximately to constellations they saw there (which have long since shifted eastward owing to the precession of the equinoxes).

Later other ancient cultures traced similar zodiac patterns in the sky. The Egyptians, Chinese and Greeks also tried to draw up terrestrial maps based on loose scientific principles. About 150 B.C. Hipparchus, probably the finest astronomer of the Greek classical period, began grouping stars into the first catalog. Three hundred years later Ptolemy borrowed this method to create a chart of the heavens that included forty-eight constellations as well as a division of stars into grades of brilliancy.

Ptolemy's system survived for centuries and was adopted by succeeding generations of astronomers for their own purposes. Astronomers increasingly added constellations more imagined than actual in order to secure a place in history. Things had gotten fully out of hand by the late eighteenth century when a German astronomer named J. E. Bode drew up celestial maps indicating star groups with names such as Officina Typographica (the printing press) and Lochium Funis (the log line).

Eventually the official list of constellations was reduced to eighty-eight, and in 1934 the International Astronomical Union rigidly defined the boundaries of the star groups in order to eliminate the mind's penchant for imposing patterns on celestial maps.

The human eye, the first light-detecting device placed opposite the viewing end of a telescope, also had a habit of seeking out patterns where none existed. This happened to Percival Lowell of the famed Boston Brahmin family near the turn of the last century. In 1894 he established, partly through his family's funds, the Lowell Observatory near Flagstaff,

Arizona. This was a place where he would be able to make unhurried personal observations of the solar system.

Lowell spent months studying Mars, drawing by hand elaborate maps of long, straight lines he believed he had detected cutting across vast Martian plains. What were they? As far as he was concerned, there could be only one explanation: The lines were artificial canals constructed by Martians to carry water across the red planet's desolate surface. The idea was so appealing that it was difficult to shake for decades.

The Martian canals made Orson Welles's 1938 broadcast of *The War of the Worlds* totally believable and enchanted generations of schoolboys, including this one, as recently as the 1950s.[1] Lowell, though, had committed no sin. Long before and ever since Lowell's canals, astronomers have seen patterns, as many false ones as real ones. By the 1990s, little had changed. Only the instruments were more sophisticated.

The Electronic Universe

Unblinking and disinterested, the photographic plate was a big improvement over the human eye. It could also do something the most sensitive eye could do only crudely: determine the intensity of light from a distant star or galaxy, essential for the discoveries that Shapley and Hubble made. In the early 1930s, astronomy changed forever when an engineer named Karl Jansky working at the Bell Telephone Laboratories in New Jersey was assigned the job of finding out why there was a steady hiss of static in radio signals beneath the usual noise of electrical equipment and weather.

He built a new kind of rotating antenna that could pinpoint the direction from which a radio signal emanated. On a hunch Jansky also had made himself familiar with astronomy. By 1933 he figured out from his new directional antenna that the static could have only one source: the Milky Way. He wrote up the discovery. Almost nobody paid attention.*

Bell officials believed that nothing could be done about the steady stream of static from space. They ordered Jansky to drop the research. A few years later a backyard astronomer in Chicago named Grote Reber heard about the work. Using a similar kind of directional antenna he had built himself, he analyzed the signals coming from space. Then he constructed a detailed map of the most intense sources of radio waves coming from the direction of the Milky Way. Still, there was almost no interest.

---

*There was only the most minor interest in the media. *The New Yorker* magazine reported upon learning about the static from the stars: "This is the longest distance anyone ever went looking for trouble."

Astronomers eventually came around to pointing their optical telescopes in the direction from which the radio signals came. Eureka. Stars, which they could actually see, were the source of the emissions. Moreover, even galaxies could send out the signals. They rushed to examine the entire range of electromagnetic radiation. The light collected by our eyes and optical telescopes was situated near the center of this electromagnetic spectrum. At the low end, with a low frequency (and high wavelength), were Jansky's radio waves. Moving up the spectrum were microwaves, infrared (or heat) waves, visible light, ultraviolet waves, X rays and, finally, high-frequency (and low wavelength) gamma rays.*

As soon as astronomers discovered that it was as easy to collect radio waves from space with ground-based directional antenna as it was to gather light waves with a telescope, the race was on to build huge radio telescopes capable of pulling in the most distant signals. The Earth's atmosphere had relatively little effect on radio and light waves, but it effectively blocked infrared and ultraviolet light waves—fortunate for living things on earth—along with gamma rays and X rays.

Instruments that could get above the obscuring molecules of the atmosphere were sent aloft aboard balloons, high-flying planes and suborbital rockets. In 1970 the Uhuru satellite, using a primitive X ray detector, took a full survey of the sky. It found that X rays were coming in from some really exotic objects: neutron stars, white dwarfs, the remnants of old supernovae, maybe even from black holes, which were just coming into vogue in the astrophysics community.[2]

Like everything else, astronomy was profoundly affected by the digital revolution, which has changed intellectual life more than any technological innovation since the invention of movable type five centuries ago.

"The data bases that we have at our disposal today are absolutely incredible," said Donald Schneider, an astronomer at the Institute for Advanced Study in Princeton. "I've got more computing power right here at this desk than you could get in the biggest mainframe computer fifteen years ago."

Schneider was fully ensconced in the new electronic age of cosmology. Blond, handsome and wearing a T-shirt and jeans, he looked more like a college athlete than a young observational astronomer playing a big role in the crisis that had begun to dominate cosmology in the late 1980s and early 1990s. One bright autumn morning in his office over-

---

*These are all waves of energy composed of electric and magnetic fields vibrating in a sine curve at right angles to each other and to the direction of travel. All travel in free space at the speed of light. The same astronomical object usually can be detected at different points along the electromagnetic spectrum, but so-called radio galaxies emit unusually intense radio waves.

looking the regal grounds of the institute, Schneider looked somewhat ruefully at his computer monitor that was linked to a powerful computer in the basement, which itself was connected via telephone lines to other computers at observatories and academic centers around the globe.

In the 1970s astronomers began relying on new kinds of electronic technologies to analyze the visible light and other electromagnetic radiation being gathered by their telescopes and antennae. The light-sensitive chemicals in a photographic plate had served as the first photometers, devices that could measure the intensity of light.

Limited by the relatively large size of the molecules in photographic chemicals, the plates were replaced by electronic photometers. The latest, state-of-the-art variety of these were called charge-coupled devices (CCDs); light waves striking CCDs after their journey across the universe and down through the mirrors of a telescope triggered electronic signals that then could be counted by a computer with stunning accuracy. New laser scanning devices could examine the spectra of starlight on photographic plates and report back directly to a huge mainframe computer, which would then decipher the results for the observer.

By the early 1990s the new electronic technology was bringing in so much information from nearby galaxies as well as from the universe's far reaches that data analyses could no longer be done by hand. Computer monitors had irrevocably replaced the yellow writing pads of Hubble, Shapley and other observers of astronomy's earlier, statelier era.

As for the astronomer, it began to seem that he or she was but an ancillary device coupled into an increasingly self-sufficient electronic cogwheel. At least Vera Rubin no longer had to spend hours stooped over a photographic plate counting the spectrum lines of her moving galaxies and then calculating the results by hand.

"The result of all this electronic wizardry is that observation is now at the leading edge of cosmology rather than theory, as it was a few years ago," said Schneider, contemplating his flickering monitor attached to a computer that ultimately was plugged into intergalactic space. "There's no way that this is going to change anytime soon."

Schneider believed that he and other observational astronomers were gathering data about so many unexpected phenomena and in such vast quantities that it had become all but impossible for theoreticians to devise mathematical formulations about these observations before the newest batch of data would pour out of the sky and into the computers. Part of the problem was that astrophysicists went about their jobs in a highly indirect manner, a modus operandi that would be unsettling for anyone expecting to get direct results.

The method was reminiscent of the dialectical method invented by Georg Wilhelm Friedrich Hegel. A thesis was created that was then opposed by its antithesis to create a synthesis, which then became a new thesis. Hegel believed that history would proceed along such a pathway to a final absolute. The work of astrophysics seems to progress along similar lines. Observers discover new objects in the sky, theorists then create a new theory to explain how the new objects fit together with the old ones; the new theory then points the observers in the right direction to look for still new objects, and so on.

However, this similarity was only superficial. Hegel's dialectic had a final goal. Science is not supposed to work teleologically toward an absolute, although many scientists have lost sight of that distinction as they search for the final great theory, the absolute truth, about the universe.

During the 1980s another problem had appeared. If Schneider was right, the cycle of observation-theory-new observation-new theory had started breaking down. Theorists could no longer coherently explain all the new observations. There were simply too many of them.

A Big Attraction

The astronomical community had sharply rebuked Vera Rubin and Kent Ford for making poor astronomical observations, then jumping to conclusions. Within two years most other astronomers were wondering why they had been so quick to criticize.

The background radiation that had been discovered in 1965 and which was considered Exhibit A for the big bang was the crucial piece of evidence on behalf of Rubin and Ford. In 1977 scientists sent balloons aloft equipped with instruments capable of detecting minute variations in this radiation. These were the most sophisticated measuring devices that had been used on the microwave radiation so far. The results surprised almost everybody. The radiation was shifted slightly toward the red end of the spectrum on one side of the sky, and slightly toward the blue end in the other direction.

The conclusion was inescapable. The Earth and the solar system were, in fact, traveling rapidly in the direction of the blueshifted background radiation. This meant that the entire Milky Way had a peculiar motion not related to the general expansion of the universe. Calculations undertaken soon afterward showed that the Milky Way was not alone on its journey across the universe. The entire local group of about thirty galaxies was moving in the same direction. The velocity was even greater than Rubin and Ford had claimed—not 500 kilometers per second but about 600 kilometers per second (about 2 percent of the speed of light).

In modern science conclusions are almost never cut-and-dried, and it was a case of good and bad news for Rubin and Ford. The velocity they had calculated of the streaming motion, as it was now being called, was close to the velocity measured by the latest instruments. But the direction the Milky Way and the other galaxies in the local group traveled was another matter. Astronomers believed that the spectrum shifts in the microwave background were an unimpeachable frame of reference for the entire universe. These spectra indicated that the Milky Way was speeding across space in almost the opposite direction from that reported by Rubin and Ford.

This was a bittersweet victory. The Carnegie astronomers' basic idea of local motion had been confirmed, but their data were thrown out. Rubin was becoming increasingly defensive about her work and tried to salvage a little glory, nonetheless. The important point, she said, was that the streaming motion of the local group was now confirmed. What caused this unexpected motion remained to be seen. The standard theories of galactic formation certainly had not anticipated this discovery.

A few imaginative ideas were floated briefly. Could immense explosions in the very fabric of the universe itself cause the local galaxies to head off in the wrong direction? Or had our local group been given an unusually powerful primordial kick early in the history of the universe? Was there a fundamental breakdown in Newton's gravity in distant regions of the universe? Most astrophysicists dropped such ideas because they were impossible to confirm with current measuring techniques, assuming that the movement of the Milky Way and the local group could only result from the gravitational pull of an enormous concentration of mass that had yet to be discovered.

And there were some gravitational possibilities. Was it a massive black hole, an object so gravitationally dense that even light could not escape its monolithic grasp? Or could it be simpler in concept and easier to detect: A cluster of unknown galaxies large enough to pull our entire local group from its normal course in the race outward with the rest of the universe?

Early surveys had found no unusually large concentration of matter in the form of galaxies in the general direction we seemed to be headed. Using Isaac Newton's law of inverse proportion for gravitational attraction, theoreticians could easily enough calculate how great the mass must be, depending on its distance from us: If it were, say, 500 million light years away, only an enormous cluster of thousands of galaxies would exert such a pull. A few hundred massed galaxies could do the job if these galaxies were ten times closer.

Determining the direction of the Great Attractor, as astronomers were now calling the source of the mysterious pull, was another matter. Not only was the local group being pulled by this unknown mass, but it also seemed to be affected by a large cluster of galaxies in the constellation Virgo, which lay nearly perpendicular to the estimated position of the attractor.

It was an unusually daunting task for astronomers to look at photographs of the night sky in a specific direction in order to determine how much mass one cluster of galaxies or another might contain. Then they had to calculate the peculiar velocities of all the galaxies that might have an effect on our own journey through space. If they could determine the effects of all these streaming motions together, astrophysicists believed, they would certainly be able to at least point their marvelous new sky-gazing instruments in the right direction.

The race was on to find the Great Attractor, with several research groups in the hunt. The observational data started pouring in. Soon it became clear that the streaming motion of our local group was accompanied by that of almost all nearby galactic clusters such as Virgo and a huge structure known as the Hydra-Centaurus supercluster. Hydra-Centaurus, about 70 million light years away and apparently in the right direction, had itself been a candidate. But astronomers found to their surprise that Hydra-Centaurus itself was flying toward something even farther out. This could only mean that the Great Attractor was far greater than anyone had imagined.

To locate the source of the enormous gravitational pull, astronomers took redshift measurements of hundreds of galaxies that all seemed to be streaming toward the same point in the universe. This required complex observations and calculations. First the galaxy's motion in terms of the universe's expansion had to be determined according to how great its light was shifted toward the red end of the spectrum. Since its expansion speed was believed to be directly related to its distance from us, astronomers then estimated the distance. By calculating the difference between the galaxy's expansion velocity and its observed velocity, astronomers determined that what was left was the streaming speed.

In 1987 a group of seven astrophysicists, who had analyzed the streaming motions of some four hundred galaxies in our region of the universe, made an announcement that shook the world astrophysics community: Every nearby galaxy, including those in clusters and gigantic superclusters, was streaming at a rate of 600 to 700 kilometers per second toward a point in the sky that lay some 300 million light years beyond Hydra-Centaurus. Nicknamed the Seven Samurai, the group calculated the mass of this monumentally Great Attractor as that of tens upon tens of thousands of thousands of galaxies.*

But what *was* it? Two members of the original Seven Samurai, Alan

---

*Members of the group were Gary Wegner, Dartmouth College; Alan Dressler, the Carnegie Institution in Washington, D.C.; Sandra Faber, University of California at Santa Cruz; David Burstein, Arizona State University; Roger Davies, Kitt Peak National Observatory; Donald Lynden-Bell, Institute of Astronomy, Cambridge, England; and Roberto Terlevich, Royal Greenwich Observatory, England.

The Great Attractor The Milky Way, with the Earth tagging along, and other galaxies in the local group speed toward an uncertain rendezvous.

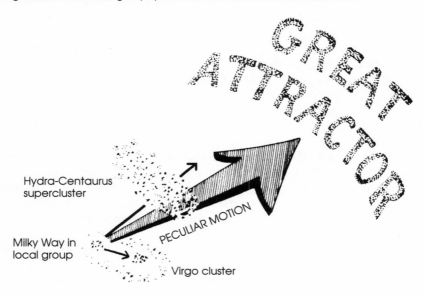

Dressler and Sandra Faber, set about to find it. Faber was a professor of astronomy at the University of California at Santa Cruz. Known for being unusually modest in taking credit for her own work, she had worked with another astrophysicist named Robert Jackson to develop a new method for determining the distances to galaxies. The method involved making measurements of spectral properties of galaxies to see how these properties changed as the galaxy's distance from us increased. The Faber-Jackson relation, as it was called, was ideally suited to the problem of pinning down the location of the Great Attractor.[3]

After two years of searching, they found it. In 1989 Faber and Dressler announced that the Great Attractor appeared to be two extremely dense superclusters of galaxies stretching 300 million light years across the universe beyond Hydra-Centaurus. From their gravitational effects, the masses of these superclusters could be calculated to be about 10,000 trillion times the mass of the sun or about 20,000 times the mass of the Milky Way. This was clearly an attractor to attract all other attractors.

After studying some 900 galaxies in the same direction, Dressler and Faber also found that galaxies such as those in Hydra-Centaurus had a much higher speed than ours, up to 1,000 kilometers per second. Then, suddenly, 150 million light years out, the peculiar velocities of the galaxies dropped to almost zero.

Those galaxies apparently had arrived at the center of attraction to take their place as little attractors themselves. And here they remained at a point some 200 million light years closer in than the Samurai had at first figured.[4]

About the same time a group of astronomers at the International School for Advanced Studies in Trieste, Italy, discovered a thick concentration of galactic structures in a region of the universe about three times the distance beyond the Great Attractor. Soon afterward a group of English and Australian astrophysicists, who had studied data from the infrared survey of the same region of sky, concluded that the Italians' supercluster probably had little gravitational effect on our local group of galaxies, but could well dominate the peculiar motions of galaxies on the far side of the Great Attractor.[5]

In the meantime, still another attractor was located. This one was in the Perseus-Pisces supercluster, in exactly the opposite direction of the Great Attractor. What could this mean? Nobody knew its mass, but a few astronomers believed it could be large enough to be tugging us backward in our race toward revelation, jubilation or only God knows what, at the center of the Great Attractor. Like a wrestler in a tag-team match, the Milky Way was being pulled in so many directions that it was hard to figure out how it all would end.

Astronomers began speculating that a proliferation of attractors must be stage-directing the comings and goings of every galaxy in the cosmic neighborhood. There were a number of questions. Was the Great Attractor itself moving in relationship to still other, greater attractors? Vera Rubin and Kent Ford had found that there were enormous quantities of nonluminous, gravitationally powerful matter within virtually every galaxy.

Was this matter distributed uniformly through the galaxies in the form of trillions upon trillions of individual subatomic particles, as many theorists had started thinking? Some thought that this kind of matter might be the gravitational seeds around which galaxies had formed. But how could such quantities of particles possibly produce the streaming motions of so many galactic clusters? Nobody had the answer.

By the early 1990s there was deep concern about the peripatetic galaxies. The unpredicted motions toward gigantic clusters of matter seemed to call into question the most prevalent and accepted theories about the evolution of the universe. Most physicists believed that matter had been nearly uniformly distributed throughout the universe shortly after the big bang. According to the standard picture, fluctuations soon appeared in this uniform density as gravitational fields of tiny clumps of matter began to attract more matter. These irregularities eventually became galaxies, then galactic clusters, then superclusters and, finally, the new class of attractors.

But cosmologists had not planned on the arrival of the attractors, which defied virtually every standard model of how the universe had evolved from the big bang. According to the prevalent theories, if there had been enough time since the big bang for galaxies to join together in groups, then they should *all* be joined in groups. That was clearly not the case.

"The problem is that if perturbations are this large on large scales, they should be even larger on smaller scales. Galaxies ought to be more clustered than they are," said Edmund Bertschinger, an astrophysicist at the Massachusetts Institute of Technology.

Before the discovery of the great galactic clumps during the late 1970s and 1980s, the universe had seemed fathomable, a place that could be understood by mortal human beings occupying a tiny planet. Now the planet, its solar system and its galaxy seemed to have set off on a journey with no known destination. It had started to look like somebody was playing a grand game with what had previously been a smoothly expanding, comprehensible universe.

If theoretical physicists were becoming anxious about the validity of their theories of cosmic evolution, they would find little comfort in the news to come.

And Then, The Wall

In the 1920s Edwin Hubble had opened the way for modern cosmic mapping when he revealed a dynamic universe filled with moving galaxies. By comparing a galaxy's redshift with its apparent faintness measured against so-called standard candles, stars of known brightness, Hubble had shown that the fastest-moving galaxies apparently were the farthest away, thus laying the groundwork for the first three-dimensional map of the cosmos.

It was not until the 1980s, with the computer power of the electronic age available to them, that astronomers were able to undertake a meaningful survey of the universe using Hubble's concept. Up until then, nobody had the slightest idea what the actual structure of the universe might be. This was when observatories were replacing photographic plates with electronic detectors, making it possible to take a galactic redshift in thirty minutes or so rather than an entire night or more. Observatories were now able to undertake regional surveys of the universe, with some surprising results. And, as already noted, instead of the uniform distribution of galaxies that they expected, astronomers began finding great clusters of galaxies, superclusters and, eventually, the immense superclusters known as attractors.

In between these gigantic new structures was another surprise: great stretches of empty space nearly devoid of any matter at all. One of these voids was an estimated 300 million light years across. This was far too immense a span of emptiness to be accommodated by existing ideas about how the universe had evolved. According to these standard theories, the cosmic density should have been as smooth as chocolate pudding.

Most astronomers simply assumed such gigantic emptinesses represented local anomalies that were meaningless in the big scheme of the universe's evolution. John Huchra, an astrophysicist at the Smithsonian Astrophysical Observatory in Cambridge, Massachusetts, was one of them.

In the early 1980s Huchra had begun working on a new redshift survey of about a third of the sky within the relatively small cosmic distance of 250 million light years. There were hints of an uneven scattering of galaxies, but Huchra was unmoved.[6] In 1985 he began working on a more systematic redshift survey that was aimed at learning how many galaxies inhabit a given volume of the universe, a long-standing problem in modern astronomy.

He joined forces for the project with Margaret Geller, a professor of astronomy at Harvard University and an astrophysicist at the Smithsonian Observatory. They began not in the observer's cage of a telescope, but in a claustrophobic third-floor room at the Center for Astrophysics in Cambridge. Stored there were hundreds of Polaroid photographs of galaxies, arranged according to their position in the sky. After deciding which galaxies they needed to study more closely, they sent the Polaroids off to an observatory on Mount Hopkins in southern Arizona. There a pair of observational astronomers named Edward Horine and James Peters aimed a 60-inch telescope at the selected targets.

New electronic devices such as image intensifiers and digital detectors allowed Horine and Peters to make measurements in a few minutes that had taken hours on a 100-inch telescope in Hubble's day. Huchra made the trip west whenever he could get in the time. He was a known telescope junkie who prided himself on his encyclopedic knowledge of the heavens.

Arizona was a long way, figuratively and literally, from the New York City area where Huchra had grown up. There it was virtually impossible to see the sky at night; the best viewing was at the planetarium in Central Park. It was not until he became an undergraduate at the Massachusetts Institute of Technology that he took a serious interest in astronomy. He often drove out to Wellesley College outside of Boston to use a telescope away from the city haze.

Deciding to make a career of astronomy, he entered the graduate program at the California Institute of Technology. He later said that he had no idea how he had been accepted into Caltech; he described himself as "scared shitless . . . intimidated from the word go." Awarded a prestigious but meager National Science Foundation grant at Caltech, Huchra needed a job to help pay the rent. He found one helping with a supernova search at the Palomar Observatory.

The rest is history, his colleagues say. Starting with his great affection for the instruments at Palomar, Huchra became one of the best hands-on telescope technicians of his generation.[7] Accustomed to wearing a great floppy Australian-style hat, he was a self-described "data merchant." He was tough to please; he would believe only that which he had seen for himself.

Huchra and Geller had a well-defined goal: to determine the redshift of each galaxy in their survey area. They did this by breaking the galaxy's light down into its constituent wavelengths with a spectrograph, then measured the resulting spectral lines to establish the redshift. Huchra loved to do the observing himself. He liked nothing better than to position himself in the observer's cage on Mount Hopkins's 60-incher for the trip out into the universe on a clear, moonless night. His personal record was taking the redshifts of 62 galaxies in one night, from dusk to dawn.

The data were collected on computer tapes at Mount Hopkins, then sent back to Massachusetts. In the beginning Huchra and Geller did not even bother to plot the redshifts on a map. There was no reason to do it. A map would reveal nothing new. In the mid-1980s everybody knew that the distribution of galaxies throughout the universe was uniformly smooth. They were only interested in seeing how many galaxies they could assay in a small slice of the sky, with the earth as its apex and a distant edge 650 million light years out.

A young graduate student from France named Valérie de Lapparent was working at the Center for Astrophysics at the time. One day somebody suggested to her almost incidentally that she make up a map of the redshifted galaxies. She did, and showed it to Huchra a few weeks later. He couldn't believe what he saw. Instead of galaxies randomly scattered throughout the survey sector, they were congregated together in enormous structures that looked like gigantic bubbles, each about 150 million light years wide.

The walls of the bubbles were formed by galaxies, and the interiors were empty. Huchra was sure something was wrong. Could they have made a mistake in their observing technology? Geller, also, had not believed earlier reports of huge cosmic structures hinted at by Vera Rubin and other astronomers.

Geller was recognized for her special talent for cosmic cartography. Her father was a solid-state physicist at Bell Laboratories in New Jersey who worked with her as child in geometrical perception. She had once considered a career in design. She had earned an undergraduate degree in physics from the University of California at Berkeley and a Ph.D. from

Princeton, which by the early 1970s was finally accepting women into its graduate and undergraduate programs.

As an adult she had retained a strong sense of geometric proportion and had what her colleagues considered an uncanny ability for three-dimensional visualization. The structures on de Lapparent's map were so obvious that she knew she was seeing for the first time a new kind of universe.

"It looked like a kitchen sink full of soap suds," she said.

In the next four years Geller and Huchra added eight additional pie-shaped sectors to their original survey. Each one brought more and more confirmation: The universe consisted of a pattern of galactic structures that utterly defied existing theory. As the data flowed in from Mount Hopkins, among the most curious of the structures that began showing up on the celestial map was an unusually large cosmic construction at least 500 million light years long and 15 million light years thick.

They could not tell its exact size because it ran off the edge of their survey, but they gave it a name anyway, the Great Wall. Geller speculated that it could be made up of the walls of still larger galactic bubbles. A system of thousands of galaxies spread across the sky in the shape of a collapsing balloon, the Great Wall was the largest coherent structure ever seen in the universe. The only thing of comparable mass that had been detected was the Great Attractor, and that still had not actually been seen by astronomers.

Theorists were stunned. This latest affront to their mathematical constructions was far too large and too massive to have formed by the mutual gravitational attraction of its member galaxies, as should have been the case. Even worse, there were indications that the Great Wall might be just one of a series of gigantic galactic sheets lined up one after the other in a honeycomb structure with voids of 400 million light years in between.

It was especially troubling that the entire Geller-Huchra survey took in less than .001 percent of the volume of the universe, an amount comparable to the size of Rhode Island compared with the rest of the Earth's surface. The vast terra incognita of the universe simply dropped off the edge of their survey like the dark abyss that seemed to swallow up most of the Earth beyond the edges of the known world on maps of the Middle Ages. Geller conceded that the coverage was small, but knew that she was seeing unexpected patterns in the distribution of galaxies.

"Something fundamental is missing in our models. We clearly do not know how to make large structures in the context of the big bang," she said.[8]

Help was on the way in the form of new instruments that would speed up the cosmic mapmaking process. One was a fiberoptic device that would make it possible to measure hundreds of redshifts each night. Assuming further improvements in technology, astronomers could expect to have accurate redshifts from as many as 2 million galaxies by about the year 2010. There was undoubtedly more news to come.

Geller and Huchra began announcing their results in 1986, first at the meeting of the American Astronomical Society in Houston, then on the astronomical lecture circuit.[9] As the year wore on, the popular media began picking up the theme—"Bubbles in the Universe," headlined *Time* magazine—assuring them a packed house wherever they went. One day Geller gave a talk at Princeton University, which, along with the nearby Institute for Advanced Study, is one of the world's leading centers for cosmology.

Her talk to a packed house of still-skeptical physicists, mostly theorists, was titled "Bubble, Bubble, Toil and Trouble." To a roar of laughter, she said she had thought about calling it *"Hubble,* Bubble, Toil and Trouble." Another thing she and John Huchra had shown was that the universe's velocity of expansion might well be substantially faster than the Hubble constant, the rate Hubble had calculated.*

Geller showed chart after chart that clearly showed the great structures of galaxies she and Huchra and found. She rhetorically wondered why no one had seen them before. Actually, she said, there had been hints of them in earlier surveys. These surveys hadn't been large enough to show the extent of the galaxies. More important, they hadn't produced anything like the map Huchra and Geller had created with the help of Valerie de Lapparent: the first three-dimensional projection of the region called the nearby universe.

There undoubtedly were skeptics left in the audience when Geller finished with her charts. A projecter whirred, and computerized 3-D projection of the galaxies began flashing on the screen. Slowly the galaxies floated by, then the entire picture shifted on its axis. It looked to the watching physicists like they were floating around the outside of a great region of space. The galactic bubbles popped in and out of view, along with the huge voids between them.

If there were any doubters still left after the light show, they were silent. Jim Peebles, an eminent theorist who had literally written the book on the large-scale structure of the universe, was in the audience. An inveterate nonreader of scientific papers, he had only heard indirectly about the new evidence of great bubbles. He had not been convinced before Geller's demonstration. "Margaret's survey convinced the last of us skeptics that there's a bubblelike nature to the distribution of galaxies," Peebles said afterward.

---

*The universe expands according to the law $v_H = H_0 R$. R is a galaxy's distance from Earth, v_H is its apparent velocity of recession and H_0 is the Hubble constant. Hubble had calculated the constant rate of expansion at about 50 kilometers per second, while Huchra and Geller and others had demonstrated that it could even double that.

The Great Wall The structure discovered by John Huchra and Margaret Geller is clearly evident in the center of the map, with individual galaxies represented by dots.

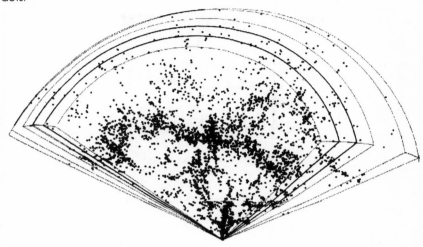

Geller's latest evidence for the bubble galaxies flashed around the astrophysics community at the speed of light. A few theorists, whose best-laid plans for a universe of even density were being dashed by findings of coherent galactic patterns, held out. However, the overwhelming amount of redshift data that Huchra and Geller had accumulated convinced most of the skeptics.[10]

British geneticist and writer J. B. S. Haldane wrote more than sixty years ago, "Now, my suspicion is that the universe is not only queerer than we suppose, but queerer than we _can_ suppose."[11]

In the intervening decades the branches of science concerned with the universe as a whole had emerged from a dark state of ignorance to one in which by 1980 theorists believed they were on the verge of understanding it all. By the early 1990s, though, Peebles (who had built a fine career on his theories about basic cosmic structure) and other theorists started conceding that the universe out there was looking wilder and wilder everyday. In fact, it was beginning to look like cosmology was about to enter another dark age.

CHAPTER **FOUR**

THE
RELUCTANT
UNIVERSE

He gave man speech,
and speech created thought,
Which is the measure of the universe.
—PERCY BYSSHE SHELLEY

If
the early findings of the Smithsonian
survey were to be believed, galaxies in our neighborhood of the universe were congregated in enormous structures such as
the Great Wall, with great voids between them. Soon after Margaret
Geller and John Huchra announced their results, however, equally persuasive evidence appeared that strongly intimated that our universe was
a completely smooth one. In such a universe matter would be distributed
evenly. There would be no place for lumps or bumps, gigantic galactic
superstructures, Great Attractors or Great Walls.

Was the universe clotted like cottage cheese or was it of an even
consistency? This conflict was to be become one of the great problems in
astrophysics in the late 1980s and into the 1990s. There were good arguments on each side.

The origins of the conflict were in the work of George Gamow, the
irascible, brilliant, practical-joking Russian émigré who had started
Vera Rubin looking into the distribution of galaxies. In the late 1940s
and early 1950s the idea of a big bang was in its infancy. Its most vocal
proponent, Gamow was especially interested in finding out how atomic
elements and their constituent parts had been formed in the earliest moments of the universe.

He was never able to develop his ideas fully about the creation of
matter in the big bang. In his day the state of nuclear and subnuclear
physics was simply too confused and overrun with unexplained subatomic particles. This, of course, didn't faze the irrepressible Gamow. A
big-idea man if ever there was one, he soon came up with the notion that

36

some heat from the big bang, hot as it had to have been, must still permeate the universe today.

He reasoned that an extremely hot big bang would have filled the universe with a primordial high-density, high-temperature radiation of extremely short wavelength. As space itself stretched out in the expansion, the energy's wavelength would have been drawn out to a greater length, and its frequency lowered.

Working with theoreticians Ralph Alpher and Robert Herman, Gamow calculated that this drawing-out-of-the-wavelength process would have shifted the initial energy of the big bang from the upper regions of the electromagnetic spectrum all the way down into the radio microwave range at the bottom. He estimated this wavelength would be several millimeters long. Gamow did some rough calculations that indicated to him that this residual energy would have a temperature of about 50 degrees Kelvin.*

Alpher and Herman immediately detected an error in Gamow's figures. Gamow was off by a factor of 10. The temperature actually would be about 5 degrees Kelvin. Gamow, Alpher and Herman believed that a perfect confirmation of the big bang theory would be the discovery of this long-wavelength radiation.[1] Almost nobody paid attention.

In those days, such a prediction seemed obscure and difficult to test. The equipment capable of such sensitive measurements didn't exist in the 1940s and 1950s. In the wake of the atomic and hydrogen bombs, most physicists concentrated their energies on trying to make sense of the infinitesimal universe within the atom and its nucleus. They forgot the notion of a resident background radiation flowing throughout the cosmos.

In the mid-1960s Arno Penzias and Robert Wilson, two young researchers at the Bell Laboratories in New Jersey, were working with a special kind of new antenna. It was built to detect the microwave radiation—with a wavelength of about 7 millimeters—from a prototype communications satellite called Echo I. As they tracked the satellite, their large, horn-shaped antenna picked up a persistent background microwave hiss. Penzias and Wilson were certain that this level of background noise was too high to be coming from the antenna itself.

Other researchers using the same antenna had ignored the hiss by simply resetting their zero, and then looked for stronger signals from space. But Penzias and Wilson were worried. Something was wrong. No

---

*The Kelvin, or absolute temperature scale, developed by William Thompson, Lord Kelvin (1824–1907) has its zero point at absolute zero (–273°C), the temperature at which all parts of a system are at the lowest energy level permitted by the laws of quantum mechanics. The Kelvin scale uses the same degree size as the Celsius scale. Thus, water freezes at 273.15 degrees Kelvin, and boils at 373.15 degrees Kelvin.

matter which way they pointed the antenna, there it was. They did everything to eliminate it; they made adjustments, cleaned the equipment, chased away pigeons. The noise still came from every direction in the sky.

One day they heard through the physics grapevine about some work at nearby Princeton being done by Robert Dicke and his student, Jim Peebles. Dicke and Peebles had, it seemed, recently theorized that the big bang should have left a background radiation of about 10 degrees Kelvin. Peebles had given a talk on a new kind of antenna Dicke proposed building to find the missing microwave background.

When Penzias and Wilson got together with the Dicke group, they suggested that they all work together. Dicke said no; he and Peebles would forge ahead on their own. When they finished their antenna, Dicke and Peebles did little more than confirm that Penzias and Wilson had indeed discovered the background radiation. The actual temperature they had found was 3 degrees Kelvin, rather than Gamow's predicted 5 degrees Kelvin, but it was close enough.

In the end both teams published their findings in the same issue of *Astrophysical Journal.*[2] Whether their finding was accidental or not, Penzias and Wilson were later awarded the Nobel Prize.

The discovery was full of irony. An unusually fine experimentalist, Dicke undoubtedly would have received a Nobel Prize himself had he decided to work with the existing antenna at Bell Labs rather than build his own. If Penzias's and Wilson's predecessors working on the antenna had not set their instruments to a false zero, they would have received the award. Moreover, Yakov Zel'dovich, an eminent physicist who was a colleague of André Sakharov and coinventor of the Soviet hydrogen bomb, saw only the early, "false-zero" papers from Bell Laboratories.

Because of the false zero, there appeared to be no background noise at all. Knowing the antenna's sensitivity, Zel'dovich, an unusually insightful theorist, concluded that the big bang theory simply had to be wrong. The greatest irony was that in 1946 Dicke himself had been a member of a group at the MIT radiation laboratory that had calculated a temperature of 20 degrees Kelvin, then collectively consigned the result to the same theoretical trash heap where the work of Gamow, Alpher and Herman was to be deposited two years later.

Big Bang or Big Bust?

The discovery of the microwave hiss quickly resolved to almost everybody's satisfaction a dispute then raging in astronomy. Did the universe begin with a cataclysmic bang or not? For nearly thirty years, the answer seemed to be yes. By the 1990s, though, the question had started to resurface.

In the early 1960s proponents of a nonevolutionary steady-state universe were on a more-or-less equal theoretical footing with backers of the big bang. The steady-state idea had been first formally proposed in 1948 by Herman Bondi, Thomas Gold and the unconventional Fred Hoyle as an alternate to the big bang.[3] During the late 1940s Hoyle became famous for a series of lectures on BBC radio that drew a nationwide audience.

During one of the British broadcasts Hoyle was struggling to explain the difference between a nonevolving and evolving universe. He had no visual aids, only the spoken word. How to describe an explosive act of creation? It was a ridiculous idea as far as he was concerned, good only for a laugh. So he gave it a ridiculous name: the big bang.

Hoyle was a little too clever for his own good. There had always been great philosophical and religious appeal for a universe without a beginning and without end. But Hoyle's idea of a static universe never caught on with the public. His name for it, the steady-state universe, may have been too mundane. In any event, the "big bang" stuck, with Hoyle opposing the very notion of it from the day he dreamed it up.

Sarcastic, smug and possessed of a peculiar accent that was a disconcerting blend of Oxbridge and, mostly, Midlands working-class, he was frequently a thorn in the side of other theoretical physicists for his pugnacious personality as well as his nonconforming views. Born into a working-class family, Hoyle had worked his way through Cambridge University. Bright and iconoclastic, he became a lecturer and then a professor, a position more exalted in the United Kingdom than in the United States.

How he rubbed his fellow physicists was not a concern of Hoyle's. He had become so famous that he would eventually be knighted by the Queen. He believed all along that the basic problem of the big bang, background radiation or not, had never been solved: that its time scale was grossly wrong because Hubble and others had erred in underestimating by magnitudes the age and size of the universe.

Moreover, he argued, the theoretical parameters for the big bang were almost infinite, meaning that the mathematics used to describe it were meaningless. In his model of the cosmos, he avoided these difficulties. In his model, matter was continuously created in a process that lead to galaxies and their constituent stars and interstellar dust. This process had produced, he said, a universe which, even though it seemed to be expanding, actually remained steady.

How did he and his colleagues account for the apparent recession of the galaxies? The trick, Hoyle said, was to create a theory in which an expanding universe always remained the same. Newborn galaxies simply take up the empty spaces left by older, departing galaxies as they rushed apart from each other in apparent recession. The universe would then always contain the same number of galaxies.

This would be true only for the universe as a whole. Any particular cosmic neighborhood could contain a different quantity of galaxies during different epochs, great periods of time that would vastly exceed the entire history of astronomical observations.[4] This allowed Hoyle and the other steady-staters to claim that astronomers simply hadn't been able to survey enough of the universe to prove the steady-state theory wrong.

Hoyle made this point over and over at astronomy conferences in a scathing, scolding voice dripping with a sarcasm that ridiculed other physicists for their belief in something as monstrously flawed as the big bang. Hoyle's theory was philosophically sounder. It required only that matter continuously be created to fill the new galaxies, not that all matter for all time be created in a single and instantaneous moment at the beginning of time.

Where did Hoyle's matter come from? According to Hoyle, it would simply pop into existence as individual hydrogen atoms, which would then coagulate into stars and galaxies. According to quantum mechanics, this was entirely possible. This system of laws explaining matter and force at the subatomic level was coming of age in the 1950s and 1960s, and Hoyle understood it better than almost any other astrophysicist.

Quantum mechanics allowed particles to materialize out of apparent nothingness in a field of space and time. Hoyle calculated the rate of particle creation necessary to maintain a steady-state universe as quite small. Only a single hydrogen atom would be required to appear per liter of space every 10^{12} years in a little burst of creation that Hoyle believed was much more likely than the big bang's single moment of all creation for all time.*

Hoyle's point was that 10^{12} years was an enormous period of time, ten times longer than the age of the universe as it was then being calculated by proponents of the various big bang models. The equations he used to describe their steady-state universe contained a modified form of Einstein's theory of general relativity, which incorporated a so-called cosmological constant. This was a disposable factor, a slight modification "necessary only for making possible a quasi-static distribution of matter, as required by the small velocities of the stars," Einstein had written.[5]

Einstein had inserted the constant to make sure that general relativity was consistent with a universe that did not change. When Hubble discovered the redshifts of galaxies twelve years later, Einstein's cosmo-

---

*In the mathematical notation used by physicists, 10^{12} is equivalent to 1 followed by twelve zeros, or 1,000,000,000,000; on the other side of the decimal, 10^{-2} would equal 0.01 or 1/100, while 10^{-12} would be the same as 0.000000000001.

logical constant could be thrown out. It had turned out, after all, that the stars did not have small velocities.

Burdened with what everybody now saw as a useless cosmological constant along with a nagging, irritating defense by Fred Hoyle, the steady-state seemed to die a natural death with the discovery in 1965 of the background radiation (although a reincarnated version appeared two decades later).[6] It was not that the steady-state could not explain the ubiquitous hiss from deep space and time. Nobody would ever underestimate the ingenuity or tenacity of Fred Hoyle. It was simply that the big bang model could so perfectly and simply account for the background radiation, an anticipated and by now necessary aspect of an evolving, changing universe.

Smooth as Cream?

Big bang backers were not yet out of the woods. A problem remained. According to Gamow's ideas as they had been worked out in detail by Alpher and Hermann, the radiation should have the characteristic of a blackbody glowing at the predicted temperature. A blackbody is one that absorbs all radiation falling upon it; it also is a good emitter of light. The spectrum of the background radiation should follow a roughly bell-shaped curve typical of energy emanating from a perfect blackbody radiator.*

In the case of the cosmic background radiation, the exact spectrum was still unknown. In the early 1970s Paul Richards, a physicist at the University of California at Berkeley, sent a balloon into the stratosphere with a sensitive microwave-detecting instrument aboard. It reported back that the background radiation had the pure blackbody thermal spectrum that would be expected from the big bang.

More accurate experiments aboard balloons and aircraft during the 1980s confirmed within a part or two per thousand that the radiation was of uniform intensity from every direction, as would have occurred if the radiation permeated the universe evenly from all directions.

The evidence seemed overpowering. Barely perceptible heat from a dying ember suggests an earlier fire; the dim glow of background radiation seemed to confirm once and for all that we lived in a universe that had arisen from a ball of intense energy.

However, by 1987 somebody seemed to be throwing clods of dirt on the big bang's primal fire. Scientists from the University of California at

---

*The spectrum of emitted radiation depends on its temperature. A high-temperature blackbody has a higher intensity at all wavelengths and its peak emission has a shorter wavelength than a low-temperature blackbody.

Berkeley and Nagoya University in Japan launched a spectrometer into orbit aboard a rocket. Its measurements showed that the background radiation was at least 10 percent stronger than the blackbody curve at several wavelengths. This was news, indeed, but was it good or bad?

Theorists who had been comfortable in their knowledge of an intact big bang theory suddenly were worried. Some pondered that the excess radiation could be the relic of an early burst of energy that had rocked the entire universe shortly after the big bang. The cosmological community was abuzz with speculation.

Could something be entirely wrong with the big bang theory? If so, what else could explain the universe as astrophysicists had been depicting it ever since Fred Hoyle's steady-state had been buried in the graveyard of forgotten theories two decades before? The worries and fears persisted for two more years.

On November 18, 1989, a day eagerly anticipated by physicists around the globe, the National Aeronautics and Space Administration launched a Delta rocket from the Western Space and Missile Center in southern California. Called the Cosmic Background Explorer, or COBE (pronounced "coby"), the satellite was about the size and weight of a large car; it was situated in an orbit 560 miles high encircling both poles and arranged to follow the moving border between night and day.

With the satellite's highly sophisticated equipment, physicists hoped by a careful examination of the background radiation to be able to look all the way back to the very dawn of the universe. To accomplish this feat, COBE carried three kinds of sensitive detectors, each designed to examine a different aspect of the background radiation from different parts of the sky. One would seek to discover minuscule variations across space, or anisotropies, in the microwave intensity. This observation could indicate how evenly matter was distributed through the universe when the radiation originated.

The detector was sensitive enough to measure these anisotropies, if they existed, to better than 1 part in 100,000. The data from the instrument, called a differential microwave radiometer, would allow scientists to look for the seeds of the large-scale galactic structures seen today by analyzing the radiation left over from the time during the universe's infancy when matter was created.

Another array of sixteen detectors was to measure the absolute brightness of the sky for vestiges of the earliest starlight. The observations from this diffuse infrared absolute spectrophotometer were expected to help explain the evolution of cosmic structure, still a mystery to theorists.

The third set of detectors would examine the infrared region of the spectrum in an attempt to find a previously unseen cosmic infrared background. Theorists believed this would have been left over from the instant the first matter appeared after the big bang. The infrared instruments also would check out the blackbody spectrum of the radiation that had been called into question by the Berkeley-Nagoya scientists.[7]

The astrophysics community held its collective breath. The launch was successful, a perfect orbit was achieved. But would the sensitive instruments aboard the $200 million spacecraft work in the hostile environment of space? Scientists at the Goddard Space Flight Center just outside Washington, D.C., had built the spacecraft and calibrated its instruments on the ground.

For two or three days they checked out the satellite's operating systems. Almost perfect. The Goddard researchers prepared to blow away the aperture covers that had protected the instruments during launch. If the procedure didn't work as intended, the covers could conceivably fly back into the instruments and destroy them. The button was pushed.

A nervous moment passed. Then the heavy aperture covers careered off into space. The instruments that could do so much to explain its origin were exposed to the cosmos for the first time.

COBE scientists had calibrated the instruments on the ground, but so far they had not been calibrated in space. The far infrared absolute

The COBE Spacecraft

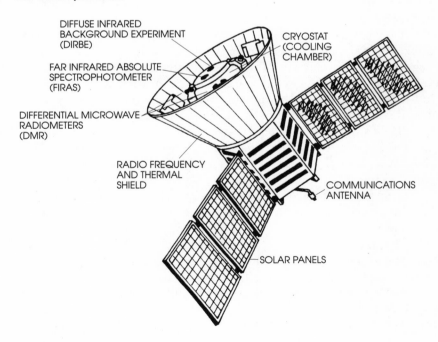

DIFFUSE INFRARED BACKGROUND EXPERIMENT (DIRBE)

FAR INFRARED ABSOLUTE SPECTROPHOTOMETER (FIRAS)

DIFFERENTIAL MICROWAVE RADIOMETERS (DMR)

RADIO FREQUENCY AND THERMAL SHIELD

CRYOSTAT (COOLING CHAMBER)

COMMUNICATIONS ANTENNA

SOLAR PANELS

spectrophotometer, or FIRAS, would check the blackbody curve, and thus confirm or deny the Berkeley-Nagoya results. The detector consisted of two horn-shaped antennae, one to collect infrared radiation from space and the other for reference.

The sky antenna looked like the bell of a trombone and was to be calibrated by a device that resembled a black trombone mute. This external calibrator, or excal, as the project scientists called it, should swing into the antenna's opening. There the excal's temperature would be regulated in order to tune the instrument for measurements of blackbody radiation.

Goddard scientists feared that the calibration procedure itself could cause irreparable harm. If its motor failed to work properly, the excal could stick in the horn, blocking all signals from space. They planned to let the instrument collect radiation on its own for a couple of weeks before attempting the calibration. In the event the excal then damaged the antenna, they would at least have some data to show for their efforts.

Scientists at Goddard monitored the satellite around the clock. Until its instruments were calibrated, however, there was little for them to do. One night around midnight three of them were watching the satellite's progress on their computer screens. They were bored.

One of them had an idea. "We've got the data pouring in," he said. "We just don't have the calibration. What we do have is the ground-based calibration. That should work just as well."

They set to work on their computers. They pulled up the original calibration done on the ground at Goddard, then fooled their computer program into thinking that this was actually the space calibration that had yet to be done. They compared this calibration with the raw data that had already started coming in from the spacecraft. After a few minutes, the monitor screen started flashing. The results were in.

"Holy shit," the three scientists said in unison.

On the monitor was an absolutely perfect blackbody curve. "It was a beautiful, smooth spectrum," Richard Isaacman recalled later. He had helped in the original calibration.

"I knew right away that it was right. I was almost positive that this was exactly the same thing we would see when we did the space calibration."

They made a printout of the screen. Another physicist, Ed Cheng of MIT, began plotting points on it by hand. Isaacman called off numbers from a computer screen. Soon Cheng hardly had to wait for the numbers; he already knew where the lines would fall. By now a small crowd of scientists from other projects was hovering over Cheng's shoulders. Nobody spoke. Cheng plotted the points, and it became more and more apparent that COBE had found a perfect fit to the blackbody curve at all wavelengths.

"I felt like I was looking God in the face," recalled Isaacman of the solemn moment.[8]

He and the others recognized immediately that the Berkeley-Nagoya results had been wrong. COBE's instruments were many times more sensitive. There could be no doubt. They seemed to be looking at the hand of creation itself at the earliest moment in the universe yet beheld by humans.

Goddard officials had sworn COBE scientists to secrecy. NASA considered the data proprietary, although they would eventually be released to the public. Also, the data were far from complete; it would take another two years to accumulate and fully analyze them all.

Although global, the physics grapevine is small and tightly wrapped. The calls started coming in: "Hey, I hear you guys knocked the bump out of the Berkeley-Nagoya stuff." "When are you going to release it?" "Are you sure you got it right?"

They were. When the instrument on board the satellite finally was calibrated more than a week later, it merely confirmed what they already knew. To a great round of appreciative applause, the preliminary results were announced at the annual meeting of the American Association for the Advancement of Science in January 1990.

Nobody knew exactly what had gone wrong with the Berkeley-Nagoya results. Perhaps the spacecraft that carried the instruments aloft had itself overheated a few errant molecules in the region of the instrument. Whatever had happened was no longer a concern.

Other COBE data, though equally spectacular, proved to be more troublesome, however. The detector designed to measure the absolute brightness of the sky, called the diffuse infrared background experiment, finished its first full survey of the sky in June 1990. Ground observers had found absolutely no variation other than from the Earth's normal motion through space in the brightness of radiation across the entire sky.

With a sensitivity capable of picking up a point in the sky that was brighter than its surroundings by as small as 1 part in 10,000, the detector was expected to find at least minuscule variations. Not a single bright spot was found anywhere. The background radiation was perfectly smooth. It radiated from all directions of the cosmos at an even 2.735 degrees Kelvin.

This was what theorists had anticipated. The eminently smooth background radiation indicated to them that in the very early universe matter had been distributed evenly. This is consistent with a violent explosion such as the big bang. And, according to COBE, the background radiation was smooth.

But there was a big problem. The background was smooth all right, but it was too smooth. It was awesomely, terrifyingly smooth. This indicated that matter had been distributed evenly throughout the very early universe. How then could a moment of creation that looked like it had taken place without a single perturbation—without bump or bruise—possibly have evolved into a universe filled with the great tapestrylike structures of galaxies that astronomers had begun to find?

Or a Clotted Cosmos, After All?

Margaret Geller and John Huchra had announced their discovery of the Great Wall the day before COBE was launched in November 1989. This concentration of galaxies 500 million light years across, 200 million light years wide and 15 million light years thick was the most gigantic construction yet discovered in the universe. Only the Great Attractor, which had been surmised but not yet seen, might be bigger. However, the Great Wall was only the beginning.

In early 1990, less than three months after the Geller-Huchra results, a team of British and American astrophysicists announced that they too had seen enormous clumps of galaxies scattered throughout the universe. In fact, it looked like the Great Wall was but one of a series of galactic sheets lined up one after the other with voids of 400 million to 800 million light years between them.

These clusters were so evenly spaced that they began to make the universe, at least those parts that had been surveyed, look like an immense honeycomb. What were the cosmologists to make of such galactic clustering?

"Such structures are unexpected in most cosmological theories," the English and American scientists wrote in the British science weekly *Nature*. "Our results are . . . possibly unappealing in terms of standard cosmogonies. It is difficult to understand . . . how so many features could maintain an organized regularity over such a long baseline."

Their discovery of the galactic honeycomb seemed to show decisively that an "inherent roughness" had been imparted to the universe almost immediately after the big bang, according to one member of the group, David C. Koo of the University of California at Santa Cruz.[9]

Yet the COBE survey had indicated, with convincing data, that there was not the slightest residual trace of a rough start in the early universe.

Such a beginning—in the form of uneven bunches of matter and energy—should have been necessary for the universe to evolve into the one filled with clusters and superclusters of galaxies now being seen.

Cosmologists had not expected and could not fathom the huge cosmic structures. They had a hard time seeing how these clusters could have evolved within the lifetime of the universe, if the seeds from which they had grown could not be found. Why would you expect to find an oak tree where nobody had ever seen an acorn?

Most theorists believed that the extremely smooth cosmological background emanating from the dawn of the universe must have somehow segregated into clumps, and then began gathering together in order to evolve into the great structures we were now finding. They thought that minuscule volumes of the early universe were slightly more dense than their surroundings. Pulled together by gravity, these tiny regions eventually grew into huge galactic formations.

Where were the minute fluctuations in the background radiation? Suddenly and dramatically in April 1992, NASA researchers led by George Smoot of the University of California at Berkeley announced that the COBE satellite had detected tiny temperature fluctuations on the order of about a hundred-thousandth of a degree in the low-level electromagnetic background radiation. According to the COBE data, the temperature was minutely higher in the direction of large galactic clusters, and slightly lower in the great cosmic voids.

The finding electrified the global astrophysics community, which proclaimed Smoot a hero. He called the discovery a "mystical experience, like a religious experience." Now it seemed possible for theorists to explain some of the structures being seen in today's universe in terms of events they believe took place billions of years ago.

Even with the COBE success, however, the big bang theory is not likely to turn out to be the final solution to the mystery of the universe's creation and structure. In the euphoria surrounding the COBE finding, it was all but forgotten that the big bang is still only a theoretical model like the others that preceded it: the earth-centered cosmos of Aristotle and Ptolemy, the sun-centered cosmology of Copernicus, the island universe theory of Immanuel Kant. And like the anomalies that eventually brought those earlier models down, problems and inconsistencies continued to confront the big bang model.

Misner's Mindbender

One of these problems had first surfaced two decades earlier. In 1969, only four years after Penzias and Wilson had made their names discovering the microwave background, Charles Misner, a young astronomer who had studied under Robert Dicke at Princeton, began wondering about the cosmic background noise.

The problem, as Misner succinctly put it, was that the radiation propagated across distances that were too great to have allowed light to travel from a point on one side of the universe to a point on the other side within the age of the universe.

Like Magellan sailing toward terra incognita, astrophysicists, riding upon feeble light waves gathered by the mirrors of their telescopes, were able to look for the edge of the universe at the very limits of observability, at what they called the horizon. If they looked at the horizon in one direction, then focused their instruments the other way, they found that the background radiation poured in from both directions at a temperature of precisely 2.735 degrees Kelvin.

Misner recognized that regions of the universe could only have reached the same temperature by having been in contact at one time. This was not a problem for Fred Hoyle's steady-state universe. Since the universe had existed forever in Hoyle's scenario, there would have been plenty of time for any two regions of the universe, no matter how far apart today, to have exchanged heat and thus homogenized.

But Misner realized this would not work in a universe that supposedly had evolved from a big bang. Most astrophysicists believed that the horizon of the universe was about 15 billion light years out in every direction. The background radiation from a distant edge of the universe had thus taken at least 15 billion years to reach Earth. The radiation from the opposite horizon also had taken 15 billion years to get here.

This meant to Misner that the two regions of space had to be separated by at least 30 billion light years. But the universe was only about 15 billion years old. How then, wondered Misner, could these widely separated regions have ever been in contact with one another in order to share the same temperature.

There simply was no way for a signal moving at the speed of light—

The Horizon Problem In this space and time diagram, regions of the universe widely separated today could never have been in contact with each other. Thus, it would have been impossible for these regions to "exchange" a common temperature at any time in the universe's history.

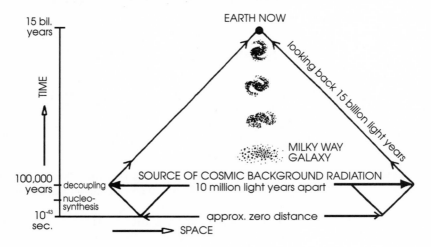

supposedly the maximum velocity attainable—ever to have traveled between the two points.[10]

As Misner worked on the horizon problem, he began realizing that things were even worse than he imagined. The evenness of the cosmic background radiation from all directions of space indicated to him that regions of the universe that are widely separated today must have had the same temperature at the time the radiation was emitted.

But it was much, much worse than he feared. When the universe was still but a toddler in cosmic terms, a mere 100,000 years old, these regions would have been separated by approximately 10 million light years—a distance 100 times that which light possibly could have traveled since the universe had begun.

Misner's horizon problem nagged at cosmologists throughout the 1970s. Like a small rock in your shoe when you are running, it was uncomfortable but too much trouble to remove. Nobody had the answer. Even after 1981, when a young theorist named Alan Guth ventured a wholly speculative supposition that seemed to eliminate the problem from the minds of many physicists, the problem remained an irritant to theorists into the 1990s. To win the marathon, you'd have to take out the rock. The horizon problem would have to be dealt with if the big bang were to be truly and fully confirmed.[11]

Cosmic Fossils

During the 1980s and into the 1990s improved astronomical observations imposed narrower and narrower limits on the vivid imaginations of cosmologists intent on explaining the origin, evolution and fate of the universe. There was Vera Rubin's missing galactic mass, along with Margaret Geller and John Huchra's Great Wall as well as the other immense new galatic configurations. Observers should have seen none of these, if the theories were right.

In the late 1980s Maarten Schmidt of Caltech, working with Donald Schneider and his Princeton colleague James Gunn, began a series of observations at the Palomar Observatory that made the work of cosmologists even more difficult.

Back in the 1960s, when astronomers had first aimed their optical telescopes at the new objects being turned up by the new radio telescopes, the big instruments at the Palomar Observatory took photographs of a few of these radio galaxies.

Strange-looking stars with an odd blue color appeared on the same photographs. Nobody had seen anything like it. Most astronomers simply assumed they were looking at unidentified stars in the Milky Way. In March 1963 Schmidt, a Dutch-American astronomer, analyzed the redshifted light of one of the blue objects. It was known as 3C 273 (or the 273rd object in the third Cambridge catalog of radio sources).[12] About the same time another astronomer, Jesse Greenstein, measured the redshift of another, 3C 48.

Their redshifts were so high it was clear that Schmidt and Greenstein had come across a new kind of celestial object. They appeared to be at an enormous distance from Earth and receding fast. Nobody knew what they were so they called them "quasi-stellar radio sources," which was quickly shortened to *quasar.*

The objects posed an array of new problems. First of all, how could they emit so much light? A powerful quasar could be a hundred times brighter than the entire Milky Way, but only about the size of our tiny solar system. Where could so much energy come from? The quasars were at the most distant reaches of the universe, some more than 10 billion light years out and moving away at the unbelievable rate of 90 percent of the speed of light.

If the redshift analyses were right, light from these quasars had taken 10 billion years to reach Earth across the depths of space. Some astronomers began thinking that they were looking out at the oldest objects in the sky, cosmic fossils of the brilliant cores of young galaxies at the dawn of time.

Two decades after he had first seen them, quasars were still his babies as far as Schmidt was concerned. Schmidt believed in the beauty and power of observation. He had few worries and little sympathy for the fears of theorists that this latest discovery or that one might not fit a certain model for the evolution of the universe.

If something didn't fit, so be it. His sartorial trademark was iconoclastic too, a bow tie that he seemed to wear everywhere. Schmidt's observing style was to concentrate on a specific area of observation, learning everything about it that he could. He was the expert on quasars.

He liked nothing more than to hop aboard the small open elevator for the five-story ride up to the observer's cage above the 200-inch mirror within the great barrel of the Hale telescope at the Palomar Observatory.*

This was the instrument that had helped him discover the first quasar, and he believed it was still the most exciting, romantic of all human tools. The colder the night, the better he liked it. He had little respect for the notion of astronomer as computer-gazer in a warm room below. He observed the stars the way they had always been seen: high up in the cold on a clear, moonless night.

Schmidt and his colleagues sought the most distant quasars. Like

---

*The instrument, for years the world's most powerful light-gathering device, was named for George Ellery Hale (1868–1938) who was largely responsible for its construction and well known for the refrain, "More light!"—a remark he probably borrowed from Goethe, who is supposed to have uttered "Mehr Licht!" as he was dying in 1832.

Cerberus at the Styx, these far objects were believed to separate the light from the primordial darkness at the edge of the universe. The only thing beyond them should in theory be the source of the damnably smooth background radiation. In the decades following Schmidt and Greenstein's discovery of them, astronomers had begun envisioning quasars as nearly mythological cosmic creatures.

Schmidt and his coterie of young observers saw some emitting enormous fountains of gas of Olympian size and speed blasting out millions of light years into space at velocities so huge they were once thought to exceed that of light. Because of their distance and energy, quasars surely would reveal much about the evolution and structure of the universe rather than obscure these secrets.

Yet finding quasars was no easy task. Dim, red objects called M stars closely resembled them. These taunted the astronomers repeatedly, with Schmidt turning up more M stars than quasars. In the early 1980s, with great patience, Schmidt and his colleagues found a number of quasars 12 billion light years from Earth. Assuming the big bang model was right, these were near the beginning of time itself, perhaps

Quasar Light   Light leaves the quasar, radiating in all directions. Billions of years later, the event is still invisible since its light has not yet reached the Milky Way. When its light finally reaches us in several more billion years, we see it as it appeared billions of years ago. The quasar may no longer even exist.

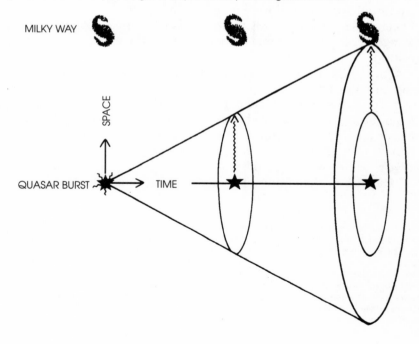

the seedlings of the earliest galaxies, or perhaps something more exotic still.

In the late 1980s Schmidt, Gunn and Schneider used Palomar's big telescope to find ten new quasars some 14 billion light years distant from Earth. One called PC 1158 + 4635 was the most distant object ever seen.* (P stands for Palomar, C is a sequence reference, and the numbers represent celestial coordinates.) Their analysis of its light indicated that the quasar could have been present when the universe was only about 18 percent of its current size and about 7 percent of its current age, or when it was only about 1 billion years old.

News of their discovery, published in November 1989, astonished astrophysicists around the globe. They were mystified. A quasar so distant and so near the dawn of time simply should not exist, because PC 1158 + 4635 left too little time in the current model of the universe's evolution to get from the big bang to stellar structures such as galaxies.

"It appears that quasars are associated with the evolution of galaxies," said Schneider. "This quasar meant that objects the size of galaxies must have formed when the universe was younger than we had thought. It puts some very serious constraints on theories of galactic and cosmological evolution." [13]

Theorists did not have an answer. Schmidt, Schneider and Gunn had begun seeing something in the universe before theoretical models of cosmic evolution said it could possibly have happened. What if a 10-million-year-old human fossil were found when all our ideas about evolution had been based on the idea that hominids first appeared just 3 million years ago? This was the same kind of question the astrophysics community faced.

"These high-redshift quasars, along with the superclusters we're seeing, are like a vise closing in on both sides of our theories," conceded Edwin Turner, a well-regarded Princeton theorist. He was shuffling through dozens of papers listing new astronomical observations, in his office one day in the autumn of 1990. "We're starting to find that we just don't have enough time to get the universe from an early state to the one that we're seeing now," he said.

"The vise is tightening. It could be about to break all the current

---

*This quasar had a redshift of 4.73, the highest yet seen. An object's redshift, called z by astronomers, is approximately the object's line-of-sight velocity divided by the speed of light. The known distribution of quasar redshifts throughout the universe peaks at about $z = 2.2$.

Theorists believe that an object with a redshift of 4.73 travels at over 93 percent of the speed of light. There is controversy among astronomers as to whether the redshifts of quasars have the same properties as the redshifts of galaxies. Some astronomers think that redshifted light has nothing to do with velocity and that another interpretation is needed.

theories. I don't have the feeling that we're working in an area we understand anymore. It's all very upsetting. We theorists definitely do get upset when we can't come up with an answer."

Turner was right. By the early 1990s the universe was in terrible shape as far as cosmological theories were concerned. As the focusing power of astronomers sharpened, frailties in cosmology's most cherished theories had become as exposed as the derriere of the emperor in his new clothes.

Just what kind of universe did we inhabit? Whatever the universe was, it was certainly no longer a consistent one.

According to the best minds in the astrophysics community, our universe was one filled with background radiation that flowed in from too great a volume of space, which itself was filled with galactic structures too large to possibly exist, which may themselves have sprung from quasars too old for the age of a cosmos in which at least 90 percent of all matter supposed to be there had never been seen.

CHAPTER FIVE

THE
PAST
IMPERFECT

*I don't pretend
to understand the universe.
It's a great deal bigger than I am.*
—THOMAS CARLYLE

The
big bang was the most recent and
most scientific of the creation mythologies. It and the
others throughout history played out a similar theme. A time before time.
A total darkness before light. Something that was uncreated before there
was something that was the universe.

In ancient Egypt, the divine he-she serpent Atum lived in the black
waters of Nun, the primordial ocean in which lay the seeds of all creation. Atum was lonely, so he masturbated to create the first creatures,
the male Shu and the female Tefnut, and gave them *ka,* the vital essence.
In ancient India the lingam (phallus) and its female counterpart, yoni,
became Father Heaven and Mother Sky.

In Polynesia the sea god Tangaroa lived alone above a vast expanse
of water. He threw down a stone. It became land. He sent a bird to plant
a vine, but it rotted. As it decomposed the maggots became people. In
Greece there was a void called Chaos. Gaea, mother of creation, emerged
from the darkness of Chaos and founded the dynastic order of the gods
who would rule the world from Olympus.

The mythologies were all strikingly familiar, haunting projections
of the human mind trying to carve out order from the bleak, random and
lonely world in which it found itself.

The big bang was the creation epic of twentieth-century physicists.
According to this scientific, mathematical mythology, at first there was
nothing—no time, no space, not even emptiness, only a void beyond
voids, a place that was no place, without color, without shape or sub-

54

stance, without the passing of a single moment or the prospect of eternity.

From this pure nothingness sprang a speck of chaos, a seed seething with such raw energy that the thought capable of contemplating it has not yet been formed. Within this speck of vibrating energy, still many times tinier than an atom, were the dimensions of time and space, although these concepts were then meaningless. There was no _now, then_ or _will be;_ no _here_ or _there._

The infinitesimal cosmos began to expand. As it grew, it cooled and its energy dissipated. Almost at once one of the forces raging within it separated from the rest. Soon pairs of particles capable of existing only in such extreme conditions flashed in and out of existence, colliding with each other in a shower of annihilation.

Suddenly the infant cosmos erupted from subatomic proportions to the size of a cantaloupe. Within a second it was as big as the solar system, a crucible of matter and energy denser than a star. Pure, energized light blazed throughout the young universe. As it grew and cooled, flying particles began coalescing into larger structures of hydrogen atoms, which eventually swirled into immense clouds of billowing gas.

Epochs passed. The universe expanded. Its blazing light faded into darkness. A million years, then a billion, came and went. All at once millions of stars began emerging from swirling clouds of hydrogen gas. Galaxies appeared, then more stars, other worlds perhaps, Earth, life, people.

The Genesis Myth

The big bang theory was so consistent with the biblically depicted beginning of the universe that a number of religious leaders jumped on its bandwagon even before scientists were able to confirm it with the discovery of the background radiation in 1965.

"True science to an ever increasing degree discovers God as though God were waiting behind each door opened by science," wrote Pope Pius XII in 1951 after hearing about the big bang.

After all, he believed in a world that also had formed from nothingness, according to the second verse of the first chapter of Genesis: "And the earth was without form and void; and darkness was upon the face of the deep." Physicists also pondered the events of the big bang, wondering whether they invoked a creator or at least some kind of a creative force.

"The odds against a universe like ours emerging out of something like the big bang are enormous," British cosmologist Stephen Hawking said during a conversation with me in 1983. "I think there clearly are religious implications whenever you start to discuss the origins of the universe. There must be religious overtones," said Hawking. "But I think most scientists prefer to shy away from the religious side of it."

Why do they shy away from it? Is it from an intellectual notion that never the twain shall meet: science and religion, distinct epistemologies with distinct regimes and different languages? Or, one wonders, is it from an unspoken fear that the big bang unconsciously arose from the biblical creation mythology, with which it seemed to share so many events and mysteries?

Creation myths invariably serve to anchor individuals in a complicated world. The mythologies also fire the imagination. For several decades the big bang has provided a fine orientation for cosmologists, serving as a road map for their best scientific instincts and creativity in the long journey back to the beginning of time. And the big bang was certainly a logically compelling mythology for a material world full of doubt and cynicism in a scientific, technological, temporal age.

Big bang proponents based their claims of scientific legitimacy mainly on the apparent recession of the galaxies, the model's explanation for the prevalence of helium and other light elements throughout the universe and the background radiation. Yet problems persisted for the scientific model of Genesis.

For one thing, it was a creation epic of ever-increasing complexity in which the so-called scientific method broke down regularly with untestable theories attempting to explain layer upon enigmatic layer of observations that seemed inconsistent with the observed "reality" of the universe. There were the galaxies that were clumpier than they should have been if the smoothness of the background radiation were to be believed. There was the fact that more than 90 percent of the matter of the universe had not been accounted for.

Physicists were able to fill in only the coarsest, most speculative outlines to explain the maelstrom of matter and force in the first instant of the universe. Nor could anyone adequately explain how the single force that was present at the creation, as the big bang model implied, had become the diverse forces seen in the universe today.

And why, for that matter, had the universe suddenly popped out of the void at all? Other than offering what at best could be called informed speculation, theorists had yet to explain what possible conditions could cause a universe such as ours to simply spring forth from nothing.

An Awesomely Flat Terrain

As well as the universe's beginning, theorists faced another big problem—its fate. Nearly a decade before Edwin Hubble discovered that the galaxies were flying away from each other, Alexander Friedmann, a mathematician at the Academy of Sciences in Petrograd (later Leningrad) became intrigued with Einstein's field equations of general relativity.

Friedmann had a sad, pale face and a great drooping moustache. He was that rare individual, a purely intellectual being, interested only in the equations' theoretical possibilities with little regard for their practical implications. He set out to solve the equations in as many ways as he could—the consequences in the real universe be damned. Friedmann was astonished at his findings. It turned out that general relativity actually predicted several different kinds of universe.

Generally, there were two kinds: one that would go on expanding forever, and one that would fall back on itself. What could this mean? Einstein himself had rejected the idea of a changing universe when he inserted the cosmological constant into general relativity. Without the constant, Friedmann showed, the universe was a dynamic entity that, regulated by gravity, was capable of growing or collapsing like a balloon.

Versatile and brilliant, Friedmann showed in a 1922 paper that the average density of matter of the universe would itself define the curvature of space. This critical mass (which Friedmann, concerned only with theory, did not bother to calculate) could easily be derived from the field equations of relativity.

According to general relativity as modified by Friedmann, we could inhabit a universe that was convex or closed, that is, space would curve back on itself the way Einstein, with general relativity, had originally described space and time. Such a universe would be finite but, paradoxically, have no boundaries. This was a geometric concept analogous to the two-dimensional surface of a sphere, except in the case of the curvature of space there were three dimensions.*

Nobody, of course, including theoretical physicists could visualize exactly what this looked like, but it was easy to express mathematically. In such a universe there would be enough matter for the gravitational force acting on it to bring the outward expansion to a halt. How much matter was required for this to occur was later calculated with great accuracy: about 3 hydrogen atoms per cubic yard of space (or 5×10^{-27} kilograms per cubic meter). Gravity acting on matter of this density, the so-called critical density, eventually would cause a universe like this to begin falling back on itself like a tire going flat.

Reversing its expanding phase, the universe would become smaller and smaller. At last, in a final gravitational atonement of ultimate and

---

*This does not mean that if you head off in one direction, even at the speed of light, that you would eventually return to your starting point, a circumnavigation that would occur on the two-dimensional surface of a globe. A universe that was closed would not last long enough for such a journey to be completed.

pure collapse, the entire universe would revert to a single point of nearly infinite density and heat.

This, of course, as many theorists speculated, could very well provide the ideal starting conditions for Big Bang II (with luck, somebody would coin a more elegant name this time around).

Or another possibility, Friedmann demonstrated, was that we dwell in a concave or open universe. Owing to the impetus impelled by the big bang, coupled with the relative weakness of gravity, such a universe would simply keep on expanding forever. This universe would be infinite in both time and space. With its curvature analogous to the horn of a trombone, it would never twist back on itself.

There simply would not be enough matter dispersed through space to allow gravity to halt the expansion of such an open universe, bringing it eventually down in a grand collapse on itself. Stars and galaxies would recede farther and farther apart in a grand and final dispersal of all matter.

There was a third possibility, too. Perhaps the universe was neither open nor closed; maybe it was balanced precariously between a fate of grand contraction and infinite expansion. In a way, this was the most difficult concept of all to grasp. Astrophysicists assigned the Greek letter omega (Ω) to represent the ratio between the actual cosmic mass density as determined by observers and the critical density that would allow gravity to pull the universe back down on itself.

If this ratio were equal to or less than 1, there would be too little actual mass density to halt the expansion, which would then go on forever. If omega were greater than 1, then the universe would be closed and the expansion would (or should already have) come to a halt. Using the example of a baseball struck by a bat and sent flying into the air, omega was similar to the ratio of gravitational energy to kinetic energy. If the batter were strong enough and hit the ball skyward with more than the critical speed represented by the ratio, the ball would escape Earth's gravity and career off into space. Fortunately for the sport of baseball (which could no longer be played if such a hitter presented himself), every batter, when he manages to hit the ball at all, always launches it at less than the critical speed.

As demonstrated countless times every baseball season, these balls reach a certain maximum height where they have an upward speed of zero, a kinetic motion of zero. Then they begin returning to Earth. Suppose, however, that the ball were struck at just the critical speed. Long after the batter had rounded the bases, showered and gone home, played out the rest of the season and eventually retired from baseball, the ball would still be traveling in orbit with gravitational and kinetic energies exactly balanced, their ratio equal to 1. An astronomer would call this a flat trajectory.

Surprisingly—astoundingly, actually—this seemed to be exactly

Gravitational Geometry In an open universe, *omega* equals less than 1, while in a closed universe, *omega* equals more than 1. In a perfectly flat universe, *omega* equals exactly 1.

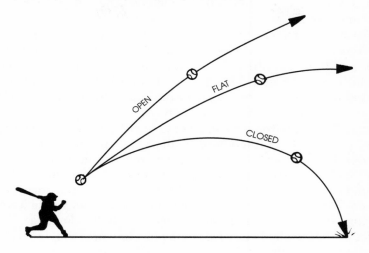

the case with the universe. Observational astronomers such as Vera Rubin and Maarten Schmidt were unable to determine whether the curved space of the universe was open or closed. The reason appeared to be that the universe was precisely, or as near precisely as anybody could tell, poised between the two states, its omega almost exactly equal to 1.

Robert Dicke, the Princeton physicist who missed out on a Nobel Prize for the discovery of the background radiation, began realizing in the late 1960s what an enormous difficulty this flat trajectory with an omega equal to 1 posed for the standard big bang model.[1] He wondered why gravity was still such a big factor in a universe that had gotten so large. Dicke believed that after 15 or so billion years that there would be at least some sign that the universe was either open or closed. Such signs did not exist. During the 1970s others such as Hawking and his colleagues C. B. Collins and Brandon Carter also began recognizing that omega = 1 caused problems for the big bang model.[2]

Theoretical supporters of the big bang believed that how the universe appears today—in terms of the number and distribution of the galaxies—had been almost wholly determined by minute features in the earliest instant of the universe. These conditions were believed to have been set when the universe was at the early age of 10^{-43} second.*

---

*This is the so-called Planck time—sometimes called the Planck wall—past which modern physicists have not yet learned to travel. Thus, it is not possible to see the initial conditions of the universe that occurred before the Planck time.

There could have been almost no deviation in conditions in the universe then to allow for the conditions we see today, just as the merest change in the angle a baseball is hit can spell the difference between a home run out of the stadium or a line-drive out.

For omega to have remained so close to 1—that is, for the universe to be so incredibly flat today—the difference between the cosmic mass density and the critical density must have been almost nonexistent in the earliest instants after the big bang. At 1 second after the big bang, it was calculated, the universe had to be fine-tuned to an accuracy of 1 part in 10^{15}, or to within 1 trillionth of 1 percent.

At 10^{-43} second, the universe would have had to have been tuned to within 1 part in 10^{59}, a fraction so small as to be incomprehensible. Had there been less matter by so much as one of these minuscule fractions, matter would have expanded outward so quickly that gravity could never have condensed the hydrogen and helium gases enough to form galaxies and stars. With just a tiny fraction more matter, gravity would have been too strong, and the expansion would have been halted long ago.

Yet in 1992 we find ourselves living in a universe that indeed seems to exist in a reasonably stable condition. What bizarre physical process possibly could have poised the universe so gracefully at the beginning for it to become what we see today?

Physicists reacted differently to this awesomely flat universe of ours. Some took the easy way out: just accept it as is and ignore the problem. During the 1970s Brandon Carter, a colleague of Hawking's, suggested this back door way to resolve the flatness problem. The earliest value of omega was simply an accidental property of the universe, according to Carter. It should be accepted as a given, with its proof apparent in the established fact of life on Earth today.

This was an aspect of what came to be called the anthropic principle. The principle seemed to call for a creator or creative force at some point in the universe's history. But it begs the question of how the universe came to be, and as a result the anthropic idea has not washed with most physicists.

"To me, it wasn't a clean argument," said John Huchra, ever the data junkie. "The universe is what it is. Just because we can't think of a good way to form galaxies doesn't mean that galaxies shouldn't be there.

"The first thing that came to my mind when I heard the argument was, okay, there is a probability that all of the air molecules in the room are up in the corner, and that's pretty small, too," he said.

"But the probability that the air molecules are distributed in this room the way they are is also pretty small. Any individual realization of a universe or a physical system, if there are lots of possible states for that physical system, has a probability that's very small. So what else is new?"[3]

Dicke and his former student, James Peebles, went after the anthropic idea with all guns loaded. In 1979 they gave the sharpest definition yet to the flatness problem, stating in a paper that the necessary initial conditions were far too special to be accidental.

"It requires very careful regulation to ensure that we do not see such things as wholesale collapse of the part of the universe appearing in the Southern hemisphere, and general expansion in the other half," the Princeton physicists stated. "It seems curious that such a small quantity should have been built into the universe at the big bang." [4]

In other words, the universe that had emerged from the big bang had a trajectory so flat that it actually defied belief. Far more understanding was obviously needed. Some physicists believed that omega must have been exactly 1 in the beginning. This, then, must still be the case today, they argued. But how so?

Since only enough mass had been detected throughout the cosmos to bring omega up to 0.1—one-tenth the required amount—huge quantities of matter were still unaccounted for, meaning that almost the entire inventory of the universe was unseen, undetected and unknown.*

---

*The missing mass needs to be distinguished from the dark matter that has been detected by gravitational studies but not actually seen. Astronomers are certain that dark matter exists. It is not certain that the missing mass exists. The missing mass has been only postulated by theorists who believe that omega must equal 1.

The visible mass in the universe makes omega equal about 0.01. Adding in the detected but unseen dark matter, which accounts for most of the universe's known mass, brings omega up to about 0.1. To make omega equal 1 would require 10 times more mass than that which has actually been seen and detected gravitationally, taken together.

CHAPTER SIX

THE
UNIVERSAL
CURE-ALL

*From the beginning that has no beginning
and the end that has no end, from the circle
of the roundness, give us your power,
Oh Great Spirit.*
—LAKOTA SIOUX RAIN DANCE PRAYER

During
the 1970s most cosmologists
stepped around the flatness problem. It simply defied
solution. Enter Alan Guth. A young theorist in elementary particle physics, Guth was working as a postdoctoral fellow at the Stanford Linear Accelerator in the fall of 1979 where he had started working on theoretical problems of particle physics relating to the big bang. A number of the theories were quite adept at predicting the number and kind of subatomic particles seen in the universe today.

Unfortunately, most had a nasty consequence: They predicted that a bizarre particle called a magnetic monopole would exist today throughout the universe in numbers at least equal to that of the proton, about 10^{80}, a staggeringly large figure. Several experiments had been launched to find the monopoles, a particle with a single magnetic charge never seen anywhere. In December Guth was thinking about how to solve the monopole problem. "I was wondering if there were any assumptions that could be changed to make theory compatible with the fact that the universe did not seem to be swimming in magnetic monopoles," he recalled.

Guth had been unable to land a permanent job and already was worried about finding a job for the next year. Late one night in his small rented house in Menlo Park, California, he sat at his desk scribbling on a pad. His wife and young son slept.

He started wondering what would have happened if the universe, instead of expanding at an even rate, had cooled rapidly right after the big bang. Suddenly he was struck with an idea so simple, so perfect and so elegant that he was sure it had to be right.

Guth was well prepared for his moment of epiphany. He had received his undergraduate and graduate degrees at M.I.T., then taught or worked as a research associate at Princeton, Columbia and Cornell. Short, always cheerful and polite, he had a shaggy mop of dark hair making him resemble Paul McCartney in *Help*.

He was known then, and still is, for being prone to fall asleep during lectures. One day in the late 1970s, the great particle physicist Steven Weinberg, author of the popular *The First Three Minutes,* which explained the origins of matter in the aftermath of the big bang, arrived on the Cornell campus. Guth did fall asleep, but he had remained awake long enough to be convinced by Weinberg that particle physicists could play a major role in untangling the mysteries of the big bang.

During this same period, Guth visited Princeton where Dicke was scheduled to give a talk. Guth had not been able to decide whether or not to go, but finally gave in. Dicke's talk was about the extraordinary flatness of the universe, the unbelievable balance of the cosmos between runaway expansion and utter gravitational despair in a great collapse that already should have occurred. Guth was fascinated. He tucked the problem away.

Wondering that night in California what could have caused the universe to expand quickly for an instant, Guth suddenly realized that the entire infant cosmos could have slipped into an unstable state that physicists call a false vacuum. Page after page, he drew figures and wrote numbers. He realized that this momentary false vacuum, occurring during a sliver of a second, would have a profound consequence for the evolution of the universe.

Guth was aware that, according to the theories of particle physics, the universe had experienced a rapid change called a phase transition as it cooled in the instant after the big bang. When water was chilled very rapidly, it could remain liquid far below its freezing point of 0 degrees Celsius. Then it would freeze all at once.

This was what must have happened right after the big bang, Guth realized. As the universe cooled, the instantaneous false vacuum created by supercooling had driven the expansion.

Suddenly Guth understood everything. It was so simple. It had to be right. No outside force, no hand of God, no divine creative power was necessary. The universe had done it all by itself.

After a sleepless night, Guth raced on his bicycle to work the next morning. There he began work on his idea in earnest. Soon he developed the equations that gave mathematical expression to his thoughts. His new version of the infant universe was startling and revolutionary, a process that he later called inflation.

Inflation should have occurred during the very early history of the universe. He calculated that this would have begun precisely at 10^{-35} second following the big bang. That was when the hyperdense conditions of the universe would have created the false vacuum condition. According to the field equations of general relativity, a kind of antigravitational force would have pushed matter apart instead of drawing it together.

This would have happened so quickly that it could only be grasped mathematically. Guth figured that within the infinitesimal span of 10^{-32} second, the antigravitational repulsion would have made the universe expand in size by a factor of 10^{50}—equivalent to doubling in size 150 times.

Far smaller than the nucleus of an atom when it began, the universe would have inflated to a diameter of 10 centimeters, about as big as a softball. This was equivalent to something the size of a grain of sand growing bigger than our universe in the same span of time.

Guth realized that this stunning instant was, if it had actually occurred, the most critical period in the entire history of the universe. When it began there had existed only the potential for mass and energy. The inflationary period ended, Guth calculated, when its inherent instability caught up with it. Guth's universe then reverted to the more leisurely pace of expansion of the standard big bang model.

Inflation was as if a colossal force had slapped the universe so hard that it changed from one size and shape to another almost at once. Except, in the case of the universe, it had slapped itself into shape.

As he worked on the equations of the inflationary epoch, Guth and his colleagues recognized they would comfortably resolve most of the anomalies then troubling cosmologists. The monopole problem was quickly eliminated: the new inflationary model predicted that the universe would not be flooded after all with trillions upon trillions of these extraneous particles. Convenient for cosmologists, according to the inflation theory, only a single monopole, possibly one or two more, would exist in the entire universe today.*

Another wonderful benefit was that inflation seemed to make the universe a natural phenomenon again. The universe was, after all, a place subject to the laws of nature rather than to a miraculous series of divinely inspired starting events. Moreover, if the theory were correct, the horizon problem no longer existed.

Before inflation began at 10^{-35} second, all the regions of the uni-

---

*In 1982 a young experimentalist named Blas Cabrera startled the world by announcing that he had found a magnetic monopole using a device built for that purpose in the basement of a Stanford physics building. I saw Cabrera's monopole catcher late that same year. The result was never duplicated. Had Cabrera by some vanishingly small chance briefly captured the single monopole predicted by the inflationary model?

verse that we observe today would have been in contact with one another. Within this momentary cosmic seed all the energy of the universe would have been evenly distributed before the exponential inflation of space itself.

Although the matter within the universe was constrained by the speed of light, the expansion of space itself during the inflationary epoch was not. It was as if somebody had stolen into a hospital and swooped away twins at the moment of birth. Inflation had done the same thing to once-adjacent regions of the universe, separating them an instant after birth at many times the speed of light, never to be joined again.

Guth also had recognized at once that his inflation scenario would solve Dicke's flatness problem by reducing it to a simple exercise in geometry. Whatever the curvature of space before inflation, it would have been flattened during the rapid expansion. An unbreakable balloon that is inflated and inflated has a surface that becomes flatter and flatter as it gets larger and larger.

The same thing happened to the universe as it expanded 10^{50} times in a fraction of a second. By then it was virtually indistinguishable from a flat universe, its mass very close to the critical value. The spatial curvature that could have given omega a value other than 1 had been driven out at the time of the exponential expansion.

The theory seemed profound and workable to Guth and his confidants. But a single problem soon looked like it could doom the idea. As he began working on the details, Guth discovered that the theory predicted that the rapid expansion would have occurred in a number of separate spatial bubbles in the nascent universe.

The problem was major. The transition from a supercooled state would not have occurred simultaneously throughout early space, but would have taken place at slightly different times in different regions. As these spatial bubbles expanded, they would have formed clusters. Like small bubbles joining together in a bathtub, the inflationary bubble universe eventually should have coalesced into a single large universe.

Guth worked on the problem for months but could not seem to get around the fact that the boundaries between the spatial bubbles should still be detectable today. Nothing like a bubble wall (not related to the Great Wall, discovered in 1989) ever had been hinted at by astronomical observation.

He feared for the theory. Finally in 1981, more than a year after the idea had first sparked his mind, Guth decided that if he went ahead and published the idea anyway, maybe others could help solve the problem of the damnable bubble walls.[1]

Guth's hunch paid off. The cosmology world was galvanized by the novel concept, which seemed to solve so many problems at the relatively small cost of the unseen bubble walls. One of those most excited by

The Inflationary Universe The universe expands rapidly, then proceeds at a more leisurely rate. In the process, many of cosmology's most vexing problems are wiped away.

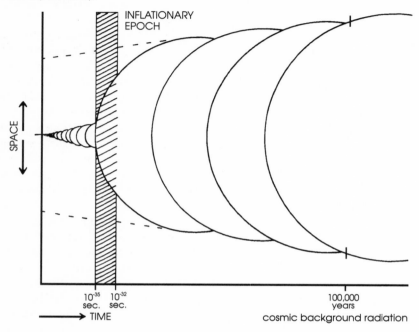

Guth's paper was Andrei Linde, a thirty-three-year-old high-energy physicist at the Lebedev Physical Institute in Moscow.

Analytical and physically athletic, even acrobatic, Linde had been awarded the Lomonosov Prize of the Academy of Sciences of the Soviet Union for brilliant work in the theory of elementary particles. Thinking there might be a solution to the bubble wall quandary in his own ideas about particle-energy interactions, Linde set to work. It was a frustrating experience. He worked for months. He could find no way to improve the situation. Nor was he willing to abandon inflation.

"I could not believe that God would miss such a good possibility to improve the creation of the universe," he said. He worked so hard without results that at one point he felt physically ill.

At last he had a breakthrough in his thinking. The phase transitions in the various regions of early space might have taken place less abruptly than in Guth's model. Linde discovered as he worked his way through the equations that the universe of today would then be free of visible bubbles and boundaries.

There would still be distinct domains, as the bubbles were now being called, but they would have developed independently of one another. A happy consequence was that in Linde's scenario the observable universe would occupy but one billion-trillionth of a single bubble domain. Its walls would be so far beyond our observational reach of about 15 billion light years that they could never possibly become visible, at least within the lifetime of the human species. Linde effectively had eliminated the walls as a major worry.[2]

This sounded highly speculative at best, more like metaphysics than physics. But it seemed to be a legitimate result, at least mathematically. Two Americans at the University of Pennsylvania, Paul Steinhardt and his student Andreas Albrecht, independently reached similar conclusions a short time later. In 1983 Linde proposed yet another hypothetical model that eliminated supercooling and the phase transition altogether while achieving similar results.

These ideas were modified and refined during ensuing years, with other theorists freely hypothesizing (along with other equally bizarre ideas) that the inflationary process could have produced trillions of universes. If so, where did that leave the human species? Surely, we had now come to the final dark circle of a long Dantesque spiral down into a sort of relativistic oblivion begun by Copernicus 450 years ago.

First we learned that the Earth was not at the center of the universe, then that there was nothing unique about the sun. We had found that our galaxy was but one among billions, and now we had discovered that our very own universe might be no place special. What dark news was still to come?

Trouble in Bubble Land

The inflationary universe made Guth a celebrity in astrophysics circles. The idea was extremely popular with theorists. In one stroke Guth had created a new vision of the big bang that was revolutionary, startling, simple, sublime. Nobody would ever view the origin and evolution of the universe the same way again.

The inflationary model so dominated astrophysics that it might have won Guth a Nobel Prize had the Swedish Academy of Science been accustomed to giving out the physics award for theoretical work in cosmology.* Guth traveled around the world to promote the idea and no longer worried about what he was going to do for work. He received

---

*In the thirty years from 1961 to 1990, there were virtually no Nobel Prizes given for purely theoretical work in cosmology, unless those were counted in which cosmology was coupled with particle physics. Presumably, this was because, unlike theories in cosmology, particle theories could easily be tested in accelerators.

nearly a dozen job offers from universities, finally settling on M.I.T. Among those impressed with the inflationary bubble scenario was Hawking.

"The idea that the universe began as a bubble offers a simple solution to many problems," he told me in 1982, a year after publication of Guth's first paper.

"Inflation accounts for the fact that the universe was fine-tuned during this period. For one thing, this allowed the universe to expand as it has without collapsing back on itself like a black hole. On the other hand, matter could also have been spread too thin for galaxies to form."

In the summer of 1982 Hawking invited Guth and several dozen other physicists to Cambridge for a series of seminars called the Nuffield Workshop on the Very Early Universe. The plan was to work out details of inflation, particularly the mechanism for seeding the galaxies.

At first the theorists were perplexed by a dilemma that seemed inherent in the concept. Inflation would have smoothed out the inhomogeneities of the earlier universe without leaving any fluctuations in the density of space capable of producing galactic structures. On the other hand, if inflation had not occurred, why was the universe so flat?

Finally, the collection of some of the finest minds in the known universe hit on the idea of applying quantum theory to the very early universe. Perhaps this would lead them out of the dark and into the promised land. This was not novel; a similar concept had been proposed nearly a decade earlier by the great Soviet cosmologist Yakov B. Zel'dovich. The problem was combining quantum mechanics, which described the behavior of matter at the atomic level, with general relativity, which explained the workings of the universe as a whole.

The two were fundamentally inconsistent. Probability reigned over the constituents of the quantum world, particles and energy. The large-scale universe, described in detail by general relativity, was a deterministic one in which starting conditions in combination with natural law would dictate the outcome: predictions of the movements of galaxies, stars and planets thus could be made with great precision. In the quantum world, the position of a particle such as an electron or photon could never be predicted precisely; all that could be said was there was a certain probability that the particle would exist some place at some time.

For more than two weeks Hawking, Guth, Steinhardt and the others scribbled long equations on blackboards, shouted at one another, and argued over dinner. They realized that the universe was smaller than an atomic nucleus before inflation began, meaning it was a region in which the rules of quantum mechanics would govern.

In such a minuscule space—one that would carry along in its expansion every aspect of the universe forever more—quantum particles flickered in and out of existence in an energy field fluctuating in intensity. Like waves bounding upon the surface of the sea, these quantum fluctuations would have peaks and troughs.[3] The peaks created by the quantum effects could easily produce density variations; an instant later inflation would stretch these fluctuations out enough to sow galactic structures. Eventually gravity would take over the job of building galaxies and clusters.

At last the combined whiz kids of cosmology reached a consensus as to the kinds of density variations that should produce the necessary contours of post–big bang space that could have led to the universe today. Working together, they envisioned a scenario that had greater detail, but was nonetheless hauntingly similar to what Zel'dovich, working alone, had produced a decade earlier.

In the years following the Nuffield meeting, dozens of theoreticians proposed scores of theories and made hundreds of predictions, all dependent upon the basic inflationary premise. Dozens of new inflationary models were created to modify, refine, supplement or supplant Guth's original.

Unfortunately, none managed to avoid a major flaw that had from the outset tarnished the sheen of inflation. Throughout the history of science the best theories always made predictions, which could then be tested by experiment or observation. Einstein's general relativity, which predicted a number of phenomena that were later observed, was a case in point. Quantum theory, envisioning numerous experiments at the subnuclear level that could then be carried out in accelerators, was another.

Unlike these theories and those in other fields of physics and elsewhere in science, Guth's and his followers' basic idea was not capable of being tested. Guth's original theory of inflation had made but a single prediction that could be considered testable: that astronomers should be able to discern the walls of domains smaller than the observable universe.

This single prediction turned out to be false. No hint of domain walls was ever observed. Linde and others simply evaded the domain-wall dilemma by creating inflationary models in which such walls could not possibly be observed from Earth. The domains, according to most of these theories, were so much bigger than the observable universe that they could never be seen, no matter how powerful the telescope. This was not the stuff of scientific testability.

The inflationary theories did seem to explain away a number of troubling characteristics of the universe—its horizon, its flatness, its clustered galaxies and so forth—all of which accounted for its enormous popularity among theorists. The reason for this apparent accuracy in explanatory power was of course simple: The inflationary theories had been created to do exactly this, explain the origin and evolution of the universe in terms that could stand up to observed detail.

In the most simple terms, the theories had been constructed in such a way that they could not be disproved.

Cold, Hot, Flat or Not?

The trend of devising elegant mathematical hypotheses which, though compelling to physicists, could not be verified was paralleled by another tendency that began making an appearance in the 1980s and 1990s: the unwillingness of theorists to give up their ideas even when contradicted by observations. One of these was a corollary to the inflationary scenarios. An important element of all these theories was the prediction that the density of matter throughout the universe would be exactly at the critical level that would make omega equal to 1.

Beginning in the 1930s, astronomers recognized that matter that they could not see with their telescopes or other instruments was scattered through the universe. In 1932 Dutch astronomer Jan Oort noticed that, as some stars began to move away from the disk of our Milky Way galaxy, they seemed to be pulled back by gravity. He added up the masses of all the stars in the disk and found there was not nearly enough mass to do the job, only about half of what was needed.

Oort made the logical assumption that the extra matter was in the form of small stars too faint for the refracting power of his telescope. The next year Caltech astronomer Fritz Zwicky noticed the same effect in a cluster of galaxies in the constellation Coma Berenices. The stars he could discern accounted for only a small fraction of the mass that seemed gravitationally evident.

By the 1970s, through the work of Vera Rubin, Kent Ford and others, it was becoming clear to astronomers that vast quantities of matter had not been seen. Much of it, though dark, was detected through gravitational analysis. All in all, 1 percent of the universe's critical mass level predicted by inflation theories was accounted for by luminous stars, interstellar dust and galaxies. By 1990 revised estimates of the masses of galaxies and clusters that included the dark, gravitational matter had pushed the total known mass inventory of the universe only up to about 10 percent of the critical level.

This was one difficulty. There was also another. Particle physicists had calculated the number and kinds of elements that should have been produced in the aftermath of the big bang. These calculations had predicted with great accuracy the proportion of light elements found throughout the universe.

Another prediction arising from these nucleosynthesis calculations was that baryons, a class of particles that included the protons and neutrons of the ordinary, everyday matter that makes up stars, planets and living beings, could account for no more than 10 percent of the universe's critical mass level.[4]

At first astronomers believed that the dark matter must consist of baryonic stuff that simply could not be seen: ordinary things like large planets, tiny stars of high density, black holes and the like. Yet virtually every inflation and nucleosynthesis calculation stated that the universe could contain only 10 percent of this kind of matter.

What, then, was the other 90 percent? Astrophysicists scrambled for a simple and plausible answer to the question. The first candidate was a particle that traveled near the speed of light called a neutrino. The wonderful thing about neutrinos—"little neutral ones," as they had been christened by the particle physicist Enrico Fermi—was that they actually existed, having been detected in particle accelerators as early as the 1950s. The bad news was that they seemed to be massless particles.*

It was also hard because of its great speed to assign the little neutrino the big job of putting a brake on the universe's expansion. Computer models showed that neutrinos probably moved too fast—close to the speed of light—to help form the cores of individual galaxies, much less to rein in the entire universe.[5]

With the apparent failure of the leading hot (for fast-moving) dark matter candidate, physicists started searching for something even more exotic. This was known as cold dark matter. It would be in the form of slower-moving, nonbaryonic particles that would have an easier time of clumping together in galactic interiors.

In the mid-1980s several physicists brought together an imaginative synthesis of ideas from particle physics and general relativity in a series of computer simulations that came to be called the cold dark matter plus inflation model. It was the creation of the so-called gang of four—Marc Davis of the University of California at Berkeley, George Efstathiou of the University of Oxford, U.K., Carlos S. Frenk of the University of Durham, U.K., and Simon D. White of the University of Arizona.

Their computer work seemed to account for galactic formation and distribution exceedingly well. Unfortunately, no kind of cold dark particle had ever been detected, although particle physicists had projected in their theories of nucleosynthesis that a number of them might exist.

One candidate was the axion, a light particle that had been invented to solve some problems of nucleosynthesis. Another was the light supersymmetric particle, an extension of a theory called supersymmetry that tried to assimilate general relativity with quantum mechanics.

---

*By 1991 some evidence was beginning to appear that a certain class of neutrino does have mass.

Inconveniently for its originators, the cold dark matter model was confronted by cold, hard and direct observation. A growing number of observations had begun showing that the universe consisted of immense structures. Margaret Geller and John Huchra had found that the galaxies tended to cluster in enormous sheets and along gigantic filaments. The Seven Samurai had discovered that our home galaxy and its neighbors were all falling toward what looked like it would be a greater mass still, the Great Attractor.

There were also the far quasars of Maarten Schmidt, Don Schneider, James Gunn and others to be accounted for. Their extreme distance seemed to indicate that stars had begun coalescing in galaxies much sooner in the aftermath of the inflationary big bang than anyone thought possible.

Geller and other observational astronomers had begun saying that the cold dark matter model was dead. Neither it nor any other theoretical construct allowed gravity enough time to create the enormous structures, such as the Great Wall, out of density fluctuations in the early universe that were consistent with a perfectly even background radiation.

"We clearly do not know how to make large structures in the context of the big bang," said Geller.

Vera Rubin was also a doubter. At the 1991 AAAS meeting in Washington, Rubin observed that *The New York Times* (January 15, 1991) and other media were publishing reports about the confused state of cosmology.

"The observers are much less confounded than the theorists," she said during her talk. "The observations are not in question; it's the theories that are."

Even inflationary guru Guth jumped on cold dark matter at the AAAS meeting. "It clearly needs to be modified for a better fit," he said.

A number of other theorists disdained the evidence. But it didn't seem to matter. Cold dark matter's coup de grace may already have been administered. The month before the AAAS meeting a group of British and Canadian astrophysicists reported in *Nature* that a new survey of the universe by the Infrared Astronomy satellite had revealed "more structure on large scales than is predicted by the standard cold dark matter theory of galaxy formation."[6] Among the coauthors were Frenk and Efstathiou, two of the theory's creators and longtime supporters. Efstathiou was especially pessimistic about the theory's chances.

"We found evidence of clustering on scales at least twice as large as those predicted by the cold dark matter model," Efstathiou said even before the report was published. This was equivalent, some astronomers thought, to the pope's converting to Judaism.[7]

The times were the most confused of any in the collective memory of the theoretical community.* Theorists struggled to develop new ideas that could explain the mounds of new data pouring in. But they seemed trapped in an extremely clumpy universe that had started out far too smoothly, even despite the miracle of inflation. Jim Peebles, Dicke's former student, had become a sort of father confessor to the astrophysics community. He was plainly worried.

"Indeed I sometimes have the feeling of taking part in a vaudeville skit: You want a tuck in the waist? We'll take a tuck. You want massive weakly interacting particles? We have a full rack. You want an effective potential for inflation with a shallow slope? We have several possibilities," he wrote.

"This is a lot of activity to be fed by the thin gruel of theory and negative observational results, with no prediction and experimental verification of the sort that, according to the usual rules of evidence in physics, would lead us to think we are on the right track."[8]

Guth's inflationary theory had easily solved many big problems in cosmology, but without bothering to make any predictions that could be tested. Perhaps inflation had been too good an idea after all.

---

*To confound theoretical matters even more, preliminary data from the Hubble Space Telescope in January 1992 indicated that the density of ordinary baryonic matter in the universe was insufficient to halt its expansion. Thus, the universe may not be closed nor balanced precisely with omega = 1 but open with an expansion that will go on eternally.

PART II

STARDUST MEMORIES

Patience, Patience,
Patience dans l'azur
Chaque atome de silence
*Est la chance d'un fruit muir.**
—PAUL VALÉRY

We had succeeded in showing that
everything was made of stardust.
We were stardust.
—HANS BETHE

▼

*Patience, patience, / Patience in the blue
sky / Every atom of silence / Is the
chance of a ripe fruit.

CHAPTER SEVEN

THE
New
Cosmology

For in and out, above, about, below,
'Tis nothing but a magic shadow show
Played in a box whose candle is the Sun
'Round which we phantom figures come and go.
—ARTHUR EDDINGTON

Jeremiah
Ostriker was at the wheel rolling
along through the modest hills of central New Jersey.
Though early March, the day was clear and bright. Ostriker's wife, Alicia, was in the passenger seat, reading *The New York Times* aloud. An article caught her eye. Astronomers at Cambridge University had made a startling new discovery. Was Jerry interested in hearing the details?

You bet he was. It was 1968 and Ostriker, then thirty and a junior member of the Astrophysics Department at Princeton, already had published several leading papers on the subject of stellar evolution and rotating stars. This sounded like it was right up his alley. According to the piece in the *Times,* the English astronomers had found strong and regular pulses of radio emissions coming from the direction of the Vela constellation in the Gum Nebula. The exceedingly regular bursts seemed to come from four pointlike sources.

Ostriker became more and more agitated as he heard the details. The fastest burst occurred exactly every quarter of a second, and the slowest every 1.3 seconds. The emissions were so regularly spaced in time, though not necessarily in strength, that the astronomers had speculated that the radio noise could be used as a clock that would be accurate to within one part per one hundred million, every bit as good as some of the best atomic clocks on Earth.[1]

Still unknown, the celestial sources of the strange radio emissions had been hastily christened: They were called pulsars, short for pulsating radio sources. According to the *Times* article the Cambridge discov-

77

erers hypothesized that the radio bursts came from some kind of previously unknown stars that were expanding and contracting in regular pulsations.*

Alicia read on. At last Ostriker, normally soft-spoken and even-keeled, could take it no longer.

"This is idiotic," he burst out.

The New Cosmologist

Ostriker was the man to know better. He recognized from his earlier work that there were several things that could cause pulsing radio bursts. A pulsating star was not among them. Already an acknowledged leader on rotating stars, he decided then and there that the regular bursts of radio noise could only be caused by something related to the rapid rotation of a stellar object of some sort.

Later that year he published his own conclusions about pulsars. His model of a pulsar called for a rapidly spinning celestial object that would be extremely compact, in fact much smaller than the Earth, probably just a few miles across. This first model of Ostriker's became the standard by which all later pulsar models were judged.

Within a few years astronomers generally agreed that pulsars were associated with neutron stars. These were compact, rotating stellar objects that might be three times as massive as the sun, but just a few miles across. Neutron stars were first postulated in the 1930s as collapsing stars too massive to become stable white dwarfs, the collapsed remnants of stars about the size of the sun.

Instead, impelled by monstrous gravitational forces, they would collapse further with a terrific release of energy, very likely in the form of a supernova, a cataclysmic explosion resulting from a star's internal imbalances. Most of its mass would be blown off into the void; this would leave the tiny compressed core with such an intense field of gravity that its protons and electrons would be forced to combine together to form neutrons, a process that would account for the radio wave emissions.

Jerry Ostriker was among the best of a new breed of what could be called the new cosmologists. The old cosmology of a less hurried, more

---

*The pulsars were first detected by graduate student Jocelyn Bell in the summer of 1967, confirmed in the fall, but not announced until the following February. The Cambridge group, headed by Antony Hewish, said they first suspected that the pulses were radio signals sent by an extraterrestrial civilization.

Other researchers accused them of ruthlessly suppressing a discovery on the pretense of not terrifying the world with the possibility of alien creatures. In reality, it was charged, the Cambridge astronomers merely wanted to ensure themselves full credit for the discovery.

elegant era had concerned itself with a more stately universe. This was the period from the 1920s through the 1950s when astronomers worried about the velocity of galactic recession, the Friedmann models, the density and large-scale structure of the universe, and the brightness of certain essential stars and galaxies known as standard candles. Their luminosity was so consistently well established that they could be used to measure other celestial objects.

This was the cosmology of Hubble and his generation of astronomers, whose main worries focused on the size and the age of the universe. In more recent times these problems were left to Hubble's anointed successor, Allan Sandage, and a coterie of his followers working tirelessly for decades at the big California telescopes, all of whom were growing old chasing the dream of cosmic structure.[2]

As for the structure _within_ the universe, the galactic and stellar stuff and all the weird objects such as pulsars and black holes and interstellar gas and invisible particles that formed the backbone of the cosmos, there was all but no concern before the 1960s. They were not the subject of conferences; no papers were written; very little work was done.

Galaxies, stars and clusters of galaxies were interesting only insofar as they could be used as measuring devices, in the form of the standard candles, for the vast spans of the universe. Nobody bothered to look out there and ask, "My God, where did these things come from?"

Had there been a blind spot? Were the big problems of the general velocity of galactic recession (or the Hubble flow as it was called) and the value of omega simply easier to formulate and, presumably, to solve (although this still has not proven to be the case)?* Gradually, though, astronomers began realizing that during their own scientific lifetimes a number of galaxies had changed.

"They started saying, 'Look, galaxies weren't there and they are there now. So therefore they were made in between now and then,'" Ostriker said. "If they were made in between now and then, they were changing during this interval. If they were changing within this interval, then they're not good standard candles." [3]

When they began realizing that the universe was beginning to present difficulties of interpretation far beyond the luminosity of this or that star or galaxy, astronomers started looking more closely at individual galaxies and clusters. With the arrival of Ostriker's generation of astrophysicists (a term that came into vogue about the same time), everybody suddenly was saying, "Forget about the standard candles. Good God,

---

*Hubble's law stated that recessional speed is proportional to distance for a homogeneous universe. If a galaxy moved away from the earth with a speed that precisely followed this law, it was following the Hubble flow.

we've got to find out where these black holes and these quasars and these structures came from."

The entire thrust of cosmology changed. Oh, the size and age of the universe and the velocity of the Hubble flow were still interesting problems; and there was still much debate on the many possible solutions to these problems; and the problems, as always, seemed to remain unsolved. But astrophysicists now wanted to know the origin of stars, galaxies, clusters of galaxies and the new superclusters such as the Great Attractor and the Great Wall.

All at once it was an entirely new universe out there, one that was pushing fiercely outward toward an unknown fate, changing, moving, filled with wandering galaxies and enormous galactic clusters, quasars of only speculative origin, unseen black holes and other uncertainties, all expanding together at such a rapid rate that the cosmos added to itself every four or five seconds a volume approximately equal to the Milky Way.

It was a universe filled with a constant flow of creation and annihilation, matter and antimatter, shadow matter, dark matter, axions and neutrinos, singularities, wormholes, quantum fluctuations, bubbles and domains. They were exciting times. Jerry Ostriker wanted to know all there was to know about this wildly changing place.

"Just a relatively few years ago astronomers here at Princeton and every place else looked at galaxies the way nineteenth-century astronomers looked at the stars," he said during a conversation with me at Princeton in 1982. "With the exception of the Hubble movement, galaxies were fixed objects, immutable, unchanging. Today we know better. They are extremely dynamic systems."

Exploring this new cosmos was a task for which Ostriker was ideally suited by temperament, inclination and intelligence. Easygoing, good-humored, sometimes even droll, he was a man as fully capable of undertaking surpassingly good physics himself as he was of running with a quiet and firm grace the meetings of Princeton's astrophysics department, which had more than once had been accused of cosseting its full share of cosmic prima donnas.

He could gently remind another faculty member, "You know, that really is an irrelevant question," as easily as he could dream up an outrageous new theory, which he could then later drop like a hot potato the moment he had second thoughts about it.

This last point was important to Ostriker. He thought that some of his colleagues in the cosmology business had untracked themselves, had even endangered their careers, by defending too long and too rigorously an idea that had outlived its usefulness. This was a sort of scientific Peter Principle. Somebody would come up with a fine theory, then another, maybe even a third. Then the same astrophysicist would create another theory that was just awful, one that was totally, terribly wrong.

Being clever, as these astrophysicists were, however, they could defend themselves over and over, sometimes for years or decades even when they were completely off-base. Ostriker thought this had been the case with Fred Hoyle and his steady-state universe. Ostriker made it a point not to defend himself or his theories against attack, to avoid the Hoyle syndrome of absolute intransigence in the face of absolute defeat. This would not get him anywhere, Ostriker believed. He would rather work on the problems inherent in the theory instead.

"I've always tried not to have an axe to grind," he said to me. "It's gotten some people in real trouble. You see it over and over."

In fact, it was a trait that had gotten the entire astrophysics community in trouble. Stubbornness in the face of a proven bad idea had helped bring about the crisis that cosmology faced in the 1990s. It was unfortunate that more of his colleagues did not or could not take Ostriker's relaxed approach to the cosmos.

One day while I was wandering around the little faux classical temple that houses Princeton's Department of Astrophysics, one of Ostriker's colleagues looked at me and confided, "He won't tell you, of course, but he's a damn near perfect astrophysicist. He's respected. He's intellectually cool. He knows how to run a department. He's able to keep things on track.

"Mostly, though, he's just a hell of a good theorist. And he's almost always right." This assessment was backed up in the citation for the Henry Norris Russell Lecture Prize of the American Astronomical Society, which Ostriker was awarded in 1980: "His output of provocative and innovative ideas is quite remarkable, especially since most of them turn out to be substantially correct."

The perfect astrophysicist even played the perfect game, squash, whose angles and spins could be calculated in various equations scribbled on a blackboard, if one was of such a mind. And, by God, he looked like a perfect astrophysicist, too: compact, receding hairline, spectacles, tweed jacket and yellowish brown academic soft soles; apparently so much so that once when he and his wife Alicia were walking down a street in India, a man they'd never seen before appeared out of nowhere and told him he was a mathematician. The man then pronounced that Alicia, a poet, was undoubtedly some kind of an artist.

During the early 1970s Ostriker and others began wondering about the origin and structure of galaxies. Remember that, in a sort of prelude to the new cosmology, in 1933 Fritz Zwicky had tried to estimate the mass of a cluster of galaxies orbiting around one another by measuring the strength of the gravity that was needed to keep the cluster together. To his surprise, Zwicky found that this total mass was at least twenty times the mass that could be accounted for by the luminous stars in the cluster.

Nobody, including the perspicacious Zwicky, worried much about

the findings for thirty years. There were other grander and more important problems to be solved relating to the size of the whole universe, not
just one peculiar galactic component. The problem of the missing galactic
mass, though, nagged at astronomers.

With the discovery of the background radiation in 1965, the feeling
that there really had been a big bang spread like wildfire. A number of
astronomers began turning their attention to how the universe had
evolved from nothing but an instantaneous burst of submicroscopic
space and time into the enormous arrays of galaxies and clusters that
were just then starting to be discovered.

This apparent confirmation of the big bang had led, also, to the
problem of the fate of the universe. If its creation had occurred as a whiff
of primordial breath, what would be its last gasp? Would the expansion
ever stop, or would it just go on forever? Some observational astronomers
such as Ostriker's colleague at Princeton, J. Richard Gott, had looked
closely at the evidence.

As far as Gott (whose name is German for God) was concerned,
there was no evidence of even nearly the amount of gravitational mass
needed to pull the universe back on itself. In such a scenario, which most
other astrophysicists believed in despite virtually all evidence to the con-
trary, the universe would collapse back on itself in a grand crunch. These
astrophysicists seemed desperate to believe that a cycle of creation, existence and decay, with the universe coming around full circle on itself, had
been ordained somewhere.

Ostriker and his colleague and friend at Princeton, Jim Peebles,
had calculated that all the mass needed throughout the universe to close
it down in on itself like an immense balloon with all its air suddenly
expelled was about three hydrogen atoms per cubic meter of space
throughout the universe. Princeton's grand old man of astrophysics and
the father of magnetic confinement nuclear fusion, Lyman Spitzer, had
stated this density amounted to somewhat less than the molecules of
breath from a single fly spread throughout the Empire State Building.

If such a density could be demonstrated, Peebles and Ostriker figured, the universe's grand finale would indeed come in an enormous im-
plosion, the so-called big crunch, which, to most theoriest, would be the
most aesthetically satisfying of all possible outcomes. There was, of
course, no immediate danger afoot. The collapse would not occur for at
least another 50 billion years.*

---

*In a story that has made the rounds through the astrophysics community, theorist John
Archibald Wheeler was once giving a public lecture on the fate of the universe when he
was interrupted by a woman in the audience. "How long did you say the universe would
last?" she asked. "Another 50 billion years," responded Wheeler. "Oh, thank goodness,"
the woman said. "I thought you said 50 million."

Theorists such as Ostriker and Peebles strangely, and at times emotionally, resisted Gott's ending: a universe expanding forever out into time and space like a balloon of infinite elasticity, forever blowing up and out, drawing its matter with it, spreading thinner and thinner into infinity, becoming colder and colder and colder.

"Not a very pretty picture," said Ostriker of the infinite balloon scenario.[4] Peebles, another rising star in the Princeton astrophysical galaxy in the late 1960s and early 1970s, argued that huge amounts of matter had simply escaped detection.

"The big question is how much have we missed," said Peebles. "Is it a factor of two, ten, fifty, a hundred?"[5]

Like Ostriker, Peebles fervently believed and expected that the matter was out there. Surely, it was in the form of some kind of invisible stellar objects such as massive black holes, interstellar dust or as some unexplained kind of matter that had not coagulated into anything larger than subatomic particles.

But there was very little evidence to indicate that such matter actually existed.

Unlikely Matter

The amount of matter needed to pull the universe back down seemed small by Earth standards. It was just a few atoms per cubic meter. In terms of what already had been found scattered through the vast reaches of the cosmos, however, this was an enormous amount.

In the 1970s, as in the 1990s, most of this matter was simply surmised. In 1973 Ostriker and Peebles jumped headfirst into the developing fray over the matter of the missing mass. They did this in the form of a gravitational study suggesting that there seemed to be far too little visible matter in the disks of typical galaxies rotating around their cores to keep them from flying apart or at least spreading out and occupying a greater volume of space.

They pointed out that observations of these galaxies seemed to show that some kind of unseen mass in outlying regions many thousands of light years from their centers provided ample gravitational attraction to hold them together in the observed sizes and shapes, that in fact a good portion of the mass resided far from the galactic centers in gigantic, dark halos.[6]

To reach this conclusion, the Princeton astrophysicists had relied on the physics of Isaac Newton and Kepler's laws of planetary motion, which state that the farther a planet was from the sun, the slower its speed in orbit. Jupiter travels more slowly than the Earth, which is slower than Mercury.

Meeting often with their notepads and coffee, Ostriker and Peebles worried about the problem for another year. They started compiling observations of mass, both visible and that which was only implied by gravitational movement, for as many galaxies and clusters as they could get their hands on. Slowly, they began to realize that the total extent and mass of these galaxies was much, much greater than the visible mass.

"What we did," said Ostriker, "was show that as you go to larger and larger scales, the mass just keeps on increasing more or less linearly proportional to the radius. As you get up to mass-to-light ratios for galaxies that are comparable to what we find in the clusters, the total extent of radius and mass was probably ten times the visible one."

The next year, 1974, they wrote up the results of their calculations with another colleague, Amos Yahil.[7] In Chicago at the meeting of the American Physical Society Ostriker announced what they had found. Not only were galaxies likely to be as much as ten times larger than they appeared through optical telescopes, but they claimed as much as 90 percent or more of the mass of the entire universe was probably invisible.

A few theorists in the Princeton department cheered the news. One was Martin Schwarzschild who had done work on Andromeda two decades earlier and found that in that neighboring galaxy there seemed to be more mass than light. Elsewhere the reception was as chilly as the outer limits of an invisible galactic halo.

Some astronomers seemed almost willfully to misunderstand what Ostriker and Peebles had said in the two papers. One camp maintained that all they had shown was that most galactic mass was in the spherical outer part rather than in the flat, central disk. Another misreading was that Ostriker and Peebles had demonstrated that the stability of galaxies *required* that they be many times more massive.

Neither was the case. But the criticism was not benign; unfortunately, it was often a typical reaction of some cosmologists when their pet ideas were threatened. The criticism was more hostile than Ostriker had encountered before. Such anger seemed to have roots in either professional jealousy for the creation of a novel and powerful new analysis, or in deep-seated ideas about galactic structure that were being threatened and which, if overturned, would have profound consequences for virtually all existing theory.

In jeopardy, it seemed, was an easy and tidy way of understanding galaxies. If Peebles and Ostriker were right, everybody would have to go back to their blackboards.

At one meeting a theorist named Agris Kalnajs presented some sketchy rotational analyses for a handful of galaxies and gave a very convincing talk to the effect that the visible matter actually was able to account for all the observed motions. There seemed to be a great collective sight of relief.

"Look," everybody seemed to say, "we don't need dark matter after all. These massive halos were all wrong. We can get rid of it, and go back to work."

Kalnajs never published the results, but the damage had been done. In the late 1970s, of course, Vera Rubin was to take direct measurements of the rotational velocities of several galaxies, then estimate their masses. She found all but irrefutable evidence that in spiral galaxies there was about five times as much mass as could be accounted for by the galaxy's visible stars. In groups of galaxies orbiting around each other, there was an even greater discrepancy: about ten times as much dark matter as visible matter.

In her 1978 paper, Rubin acknowledged that her observations confirmed the 1974 work of Ostriker and Peebles. In ensuing years it was all but forgotten who had broken the ice on dark matter. By the late 1980s students occasionally would come up to Ostriker to tell him he probably was too old to realize that times had changed, that galaxies had massive halos, and so forth. Ostriker, the victim of his own success, was more amused than alarmed. He didn't care who received credit. He and Peebles had been right all along.

The universe, to paraphrase Tennyson, was half revealed, half concealed. No, even that was not exactly right: the universe was barely revealed, and all but concealed.

The Galactic Glue

Some kind of mysterious, invisible stuff apparently was keeping the galaxies and clusters intact. With this realization, the astrophysics community changed direction again.

In 1990 I spoke with John Archibald Wheeler about the state of cosmology. Wheeler was Princeton's eminent cosmologist-cum-natural philosopher who had coined the term _black hole,_ had been Niels Bohr's protégé, a friend of Einstein's, and the mentor of Richard Feynman. According to Wheeler, dark matter was the "hottest topic in physics."

But what was it? Wheeler, as always, had ideas, but didn't profess to know with any degree of certainty. Nor did anybody else. This was a perplexing and unusually frustrating state of affairs. Whatever the dark matter was, it was the most common material in the universe, and cosmologists had failed utterly to explain it.

Did the unknown matter consist of massive black holes or trillions of microscopic ones or huge numbers of unseen planets or stars too small and too faint to be observed? Or was it a known kind of subatomic particle, such as the neutrino, which had a degree of mass that had yet to be detected? Was it even an unknown kind of particle that could interact with other matter only through the force of gravity, rather than through one of the three far more powerful forces that control the interactions of the subatomic world?*

The dark matter was crucial. The evolution of the universe depended on it. By the 1990s the new clusters and superclusters, which were being discovered practically every other day, could only be explained in terms that took into consideration an understanding of the nature of this dark form of reality (if that indeed was what it was).

Moreover, this lack of knowledge could even begin putting serious constraints on the big bang itself. It had become essential to reconcile the smoothness of the cosmic background radiation with the lumpiness of matter in our locality of the universe. Otherwise, there was no way to explain why the universe had started out on such a smooth road, but had ended up on such a coarse, rutted one.

Alan Guth's a priori inflationary model, the darling of cosmology during the 1980s, had solved in a flash a number of serious difficulties in the standard big bang model—the flatness and horizon problems, to name two. A number of cosmologists even thought that inflation pointed the way toward a solution of the lumpy galaxy problem. One of the requirements of Guth's original inflation theory and its many offshoots was that the universe have a density that would be just sufficient to prevent expansion onward and outward forever.

Controlled by inflation, then, the universe would be closed, not open. This, of course, was another plus in inflation's favor as far as theorists were concerned, despite the fact that nobody had yet devised a means of actually testing the theory. Conveniently, however, inflation actually had called for a universe consisting of as much as 99 percent dark matter.

This was not exactly a prediction of inflation, though, for the model was created with the dark matter in mind. The detection of every bit of dark matter in the universe would not prove inflation. The theory had been too artfully crafted. Especially with the help especially of Andrei Linde, inflation simply avoided making any testable predictions at all.

The Totalitarian Regime of Gravity

A lovely metaphysical speculation, inflation did provide theorists with a highly useful framework. It had led, for instance, to the cold dark matter theory of galactic development, which had been formulated by the gang of four (Davis, Efstathiou, Frenk and White) in the mid-1980s. Of course, cold dark matter had yet to be identified, a problem of more than

---

*These are electromagnetism, the weak nuclear force, and the strong nuclear force.

passing concern given that this was the stuff that made up almost all of the universe.

Then there was the difficulty of the theoretical fit. With the discovery of ever greater galactic structures and the ever smoother cosmic background radiation being found by the COBE satellite, problems had arisen for the cold dark matter model. The model required constant revision to accommodate the flow of new data. More problems for cold dark matter came from an even finer resolution of the background radiation that had been obtained at an observatory at the South Pole run by a team of Princeton scientists. The result was the same: no primordial inhomogeneities.

Very distant and thus, it was surmised, very old objects also caused concern. The cold dark matter theory predicted that galaxies had developed at least several billion years after the big bang. Yet score upon score of quasars, which many astronomers speculated were infant galaxies, had been found more than 10 billion light years out, meaning they had been formed near the dawn of time.

The cold dark matter model could account for the new quasars only by a stretch. And even if the quasars were discounted altogether, there was a handful of huge new radio galaxies at distances of more than 11 billion light years.

During the late 1980s Ostriker and other theorists spent more and more of their time trying to figure out how and when gravity had begun its creation of galaxies. The main challenge of the moment was to find a theory of galaxy formation that also fit the new observations of large clusters and the superclusters. In addition, theorists needed to explain the initial positions and movements of the lumps in the primordial cosmic mass about the time of the big bang.

These lumps, astrophysicists imagined, eventually coalesced into galaxies. Of course, these ancient lumps and bumps had yet to be detected in the cosmic background noise.

The seed for modern theories relating to the formation and distribution of galaxies, including a wildly speculative one from Ostriker, had been planted in the 1930s. The perpetrator was one of the oddest characters to shuffle onto the stage of twentieth century physics, a plump, often irritating Belgian priest and mathematics professor named Georges Lemaître.

Often ahead of his time, even prophetically so, Lemaître was frequently ignored by the physics greats of his day. At one meeting when he was unusually obtrusive, even the normally kind and tolerant Einstein pushed him aside. Two years before Hubble's 1929 announcement that the galaxies were receding according to their redshifts, Lemaître published a prophetic paper in the all-but-unknown journal of a local scientific society in Brussels.

In it he drew a mathematical theory that linked the few redshifts that already had been observed with general relativity and concluded that the universe was expanding.[8] The paper was virtually ignored for more than two years. Finally, Arthur Eddington, an English theorist who was then the acknowledged grandmaster of the world physics ring, conceded that Lemaître had written a revolutionary paper.

A short time later Lemaître speculated that the universe must have expanded from an extremely dense point, which he called the primeval atom.* In 1933 the visionary theorist-priest went further still, hypothesizing that the dense mass in the early universe must have been very smooth. Here and there, however, it could have been gathered together in little clumps Lemaître believed, although he did not suggest what might have caused these. He recognized that where the density was greater the force of gravity would be stronger. This would attract nearby mass, and the gravity of the larger clump would be proportionally more powerful.

Such a process would continue, Lemaître thought, with the tiny primitive clumps of mass leading to larger and larger ones as the universe cooled and expanded. Eventually a hierarchy of structures would emerge: galaxies, clusters of galaxies and so on. Their size and location today would have been predetermined by the size and placement of the initial lumps.[9]

Lemaître's early ideas were tentative, unquantified and highly speculative. Not all, nor even most, physicists believed them. Lemaître grew old watching more celebrated physicists follow the course he had set. He lived just long enough to learn of the discovery of the microwave background radiation, which seemed to suggest that his path had been the right one.

About the same time—in the mid-1960s—Jim Peebles made the first stab at attaching numbers to Lemaître's gravitational hierarchy model.[10] This work spawned a generation of supercomputer models that attempted to simulate the growth and distribution of ripples of matter in the expanding universe. One computer model was the cold dark matter theory of galactic formation. Its creators used millions of initial mass points in a sophisticated version of Lemaître's gravitational hierarchy theory.[11] Neither it nor any of the others produced a result that accorded well with actual observations of the movement, location or distribution of galaxies. This was true, the model makers conceded, even when the models were given almost complete freedom of choice in choosing possible initial conditions.

---

*This first notice of the big bang provided grist for Fred Hoyle's scathing criticism of the expanding universe theory. What kind of scientific theory was this, Hoyle wondered later, that had been proposed by a priest and endorsed by the pope?

Two Gravitational Models   In the bottom-up gravitational hierarchy model on the left, small clumps of primordial matter gather mass to form large bunches, which eventually become galaxies. In the top-down pancake model on the right, large primordial clumps collapse into thin pancakes, which later fragment into many galaxies.

GRAVITATIONAL HIERARCHY MODEL   PANCAKE MODEL

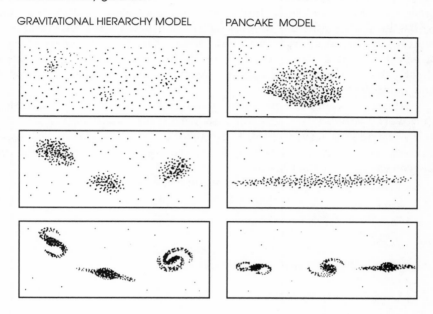

Computers, of course, weren't the only way out of the dilemma. Joseph Silk, like Stephen Hawking, had been inspired by theorist Dennis Sciama at Cambridge. Silk eschewed computer models. He preferred to work with mathematical concepts rather than dabble in contrived electronic images. In 1967, while he was working on his Ph.D. at Harvard, Silk attended an annual summer workshop on fluid dynamics at the Woods Hole Oceanographic Institute. By chance, the topic that year was the fluid-dynamical creation of galaxies in an expanding medium.

The subject, it turned out, provided an alternative way of looking at big bang cosmology. Silk began thinking that gravity might not keep initial lumps of matter in the early universe intact if they were much more than a thousand times smaller than a typical galaxy of today. He thought that the fledgling galaxies would probably break apart under the onslaught of the intense radiation then bombarding the newborn cosmos.[12] A few years later a group of Soviet theorists, led by the great Yakov Zel'dovich, extended this idea to suggest that the first clumps of matter in the early universe grew fast and large.

As the lumps expanded and cooled, they would collapse under their own weight into a thin circle of gas that looked like a pancake. These thin shells eventually would break into many pieces, becoming the galaxies lying along great two-dimensional surfaces, which Margaret Geller and John Huchra and others were discovering in the 1980s.

The Lemaître-inspired gravitational hierarchy was a bottom-up model in which minuscule lumps eventually condensed into galaxies, while the pancake model was top-down. It started with big chunks of matter that broke apart into smaller structures. If, as some of the new observations suggested, galaxies really were distributed along huge sheets, then the pancake theory just might explain the great structures now being seen out there.

However, could the pancake model account for sheets of galaxies that were exceedingly thin compared with their height and width? As Jerry Ostriker observed, all of these models, indeed even the model of models, the big bang, still assumed that gravity was the major worker in the construction of the cosmos.

Was there something else? he wondered.

CHAPTER **EIGHT**

A
PROFUSION
OF
CONFUSION

Och de har var och en sin värld och rymd,
För all andra ögon skymd,
Med egna månars makt och trolleri
*Och egna vintergator däruti.**
—BERTIL MALMBERG

Dark
matter, which seemed so intimately
bound up with the origin, structure and fate of the
universe, was an ugly picture in a lovely frame. Another physical force besides gravity could be needed to be invoked to explain the strangeness of the patterns that astronomers were seeing in our cosmic neighborhood.

Jerry Ostriker thought this probably was the case. In any event, the answer, if it ever came out at all, would be full of surprises. One day over coffee in his office he told me that he was also sure that whoever came up with the answer would be somebody with a healthy irreverence for the prevailing theory of the day.

Ostriker liked to putter around with ideas about the sociology of the physics community. He believed this might lead him to a little insight into his colleagues. Through such puttering, which he conceded was mostly anecdotal, he had concluded that North American theorists generally came from one of the two coasts, while experimental physicists mostly came from the midlands. I pointed out a couple of notable exceptions: his own theoretical colleague at Princeton, for instance, Jim Peebles, who came from Winnipeg, Manitoba.

---

*Each one has his own world and space, / hidden from all other eyes, / with the power and magic / of his own moons and galaxies.

"Ah, well," Ostriker said cheerfully, "this is only a little theory of mine."

Another idea he had was that the best scientists, wherever they came from, were practically born irreverent. He believed budding little prescientists tended to perform poorly in their earliest schooling because of a lack of interest in conforming.

"Scientists inhabit quite an inhuman world, and so they tend to believe in a universe beyond people," he said. "And young people destined to become good scientists tend to be more curious about the universe around them than about other people."

For a physicist a tree falling in the forest did indeed make a noise, and one of defined amplitude and frequency. Whether or not a person was there to hear it was absolutely irrelevant.

"Prescientists don't believe their teachers, anymore than adult scientists pay much attention to lawyers," Ostriker said.

Some Kind of Explosion?

Ostriker seemed to fill the bill for his own description of irreverent. His mind was so open that, unlike most other scientists, he didn't even dismiss astrology out of hand. He had seen real effects taking place among the planets.

"Mars always shows the same side toward the Earth whenever it's closest to us, so we seem to have altered its rotation rate," he said. "So there's one effect. In another case, radio emissions from Jupiter seem to be mediated by one of the other planets."

Ostriker's irreverence had overturned the company line on pulsars. Now he would train a different eye on galactic evolution. In the late 1970s he began scribbling on his notepads, looking for something besides gravity, but as powerful and as ubiquitous, that could play the leading role in galactic production.

Stellar evolution had always intrigued him. He had spent more time studying stars than most theorists, who as a rule were after bigger game. Might the way stars evolve over time provide a clue to the way galaxies themselves had originated? Which had formed first, galaxies or stars? Most cosmologists thought that galaxies had coalesced first, then fragmented into individual stars.

Ostriker began wondering. What if some kind of an explosion were brought into play?

Explosions did happen. Humanity had seen them throughout history in the form of supernovae, which occurred when a star underwent an internal imbalance caused by the exhaustion of its nuclear fuel. Apparently taking place in large stars with a critical mass much greater than the sun's, the imbalances would take the form of accelerated nuclear reactions. This would lead to enormous temperatures up to 5 billion degrees Celsius, followed by massive gravitational collapse.

A cataclysmic explosion would follow. The star would blow itself apart, emitting as much energy in a few seconds as the sun would emit over millions of years. The burst would momentarily light up the night sky, its magnitude equal to that of an entire galaxy. After it was over, all that would be left behind would be a filament of gas so tenuous that the atmosphere we breathe here on Earth would be millions of times denser.

Astrophysicists believed the original core might survive as a neutron star, a pulsar, or even a black hole. One of these explosion remnants was believed to be the Crab Nebula. It probably resulted from a supernova recorded in China in 1054 that was so bright it could be seen during the day.

Ostriker knew supernovae were fairly rare in today's universe, occurring as they did out of the main sequence of stellar evolution of average-size stars (of which our sun was a charter member). What if many, many supernovae had occurred in the universe's younger years? They certainly packed a wallop. Could there have been enough of them to aid in the creation of galaxies long ago? More than a decade earlier the imaginative Russian theorist Yakov Zel'dovich had wondered along similar lines and with a pair of theoretical colleagues in Moscow had written up the idea.[1]

Ostriker worked out his scenario with Lennox Cowie, a young colleague, borrowing heavily from theories of stellar evolution. These theorists believed that in the universe today shock waves from a supernova explosion could compress a cloud of gas and dust into stars. What about when the universe was much younger?

Perhaps there had been just a few flimsy galaxies—perhaps even just one—containing only a few stars that had somehow managed to gather themselves together somewhere in the early universe. Some of the stars could have burned out and exploded as supernovae at a relatively tender age. The concussion from such an explosion would blast far out into the infant universe, carrying uncondensed, primordial gas from the young galaxy with it. This could then create a huge bubble of gas in intergalactic space. Where this bubble of gas bumped into other similar bubbles, the gas would fragment and coalesce into new galaxies. These clumps of matter would be large enough to be drawn together by the force of their own gravitational mass, pulling additional intergalactic matter inward as well.

Ostriker thought it would be likely that some of the stars forged in the galactic furnace of the churning, unsettled young universe would, after a short, unstable life, burn out and explode as supernovae. Ostriker imagined that the process would occur over and over again.

"Galaxy formation would come about as a chain reaction, a series of gigantic cosmic nuclear explosions," he said.

Ostriker and Cowie published the new theory in 1981.[2] There was very little interest. Like most other theories those days, their idea was highly speculative. Besides, there was virtually no persuasive evidence for primordial explosions. And, in the early years of the 1980s, the cold dark matter model was becoming fashionable among cosmologists as the most likely perpetrator of galactic creation.

This was a model, after all, thus was straightforward and elegant, and it seemed mathematically compelling. Moreover, its creators—Davis, Efstathiou, Frenk and White—had done their computer homework and were able to present dramatic and persuasive graphics at astrophysics meetings.

Remember, though, the problems that had developed as one leading candidate after another for the cold dark matter itself was jettisoned for this failure or that one. Particle physicists had now joined the chase. Experiments at large accelerators had ruled out a number of contenders such as slow-moving, massive neutrinos. Other theoretical particles had such an awesomely high energy level that they exceeded the capacities of the current generation of accelerators.

After early reports of the Geller-Huchra survey showing large galactic structures were announced in 1986, doubts about cold dark matter began floating through the air at astrophysics conferences like filaments in a galactic nebula. The advocates still persisted, of course.

As far as he was concerned, said Carlos Frenk, the Great Wall, along with the array of thousands of galaxies shaped like a little Mayan hieroglyphic creature and all the rest, were "rubbish." Frenk and a few others still maintained that the observations encompassed such a narrow region of the sky that they provided only "anecdotal evidence" about the real structure of the universe.

"It's not a statistical analysis," he complained.

Other theorists worried about drawing permanent conclusions from the slick graphics that had been generated by the computers at the Center for Astrophysics. Jim Peebles, who apparently once had been convinced by Geller and Huchra's redshift data, was having second thoughts. He passed these on to his fellow cosmologists in the summer of 1990 at the Nobel symposium in northern Sweden on the birth and evolution of the universe. "Beware of Great Walls," he warned. "The eye is a great recognizer of patterns."[3]

Patterns in the sky or not, the cold dark matter model seemed incapable of explaining how galaxies might have evolved in terms of time and scale. In the skillful hands of Marc Davis, a Cray supercomputer could juggle as many as 250,000 clouds of theoretical dark matter. The runs on Davis's Cray, taking as long as twenty hours, simulated the accumulation of swarms of dark matter into computer galaxies about 10 billion years ago. Then in another 5 billion computer years, they would finally coagulate into clusters.

Davis's computer models were extremely adept at juggling new data to a custom fit. The astrophysics community found them highly compelling. Even so, the computer models still had trouble projecting the existence of galaxies much older than 10 billion years old and, of course, by the early 1990s such galaxies were being found in increasing numbers.

Another flaw was that the cold dark matter computer models relied on a concept called biasing, which made a number of assumptions. Among these was that actual galaxies in the universe really gave an inaccurate or biased picture of the matter within them. So the models were constructed in such a way that they could simply rearrange the distribution of dark matter within a galaxy at will. This did have the consequence of making it easier to match the models with observational data. But the adjustable variables, or free parameters as they were called, made the models' predictive powers less than satisfying.

Things got only worse for cold dark matter. In mid-1991 two new quasars were found that posed new challenges to the computer simulations. Richard McMahon of Cambridge University and Michael Irwin of the Royal Greenwich Observatory discovered one using the U.K. Schmidt Telescope trained on the southern sky from Epping, Australia. It was the brightest known object in the universe, shining with the brilliance of 10^{15} suns.

This was a startling discovery. The extreme luminosity and distance of the new quasar (officially known as BR 1202-07) seemed to indicate that some galaxies must have been present when the universe was but 7 percent of its current age, probably less than 1 billion years old. How could this be explained? The most straightforward of the cold dark matter models was utterly incapable of showing how fluctuations in primordial matter, if they existed at all, could have grown fast enough to form such a quasar.

Using the Palomar Observatory's 5-meter telescope, Schmidt and Gunn found the other quasar. It had a redshift of about 4.9, and was believed to be the most distant object of all. If the redshift measurement were right, the quasar would have existed when the universe was only about 6.7 percent of its present age.[4]

Gunn was almost apologetic for the discovery. He was fearful that the new observations would be taken the wrong way. In his opinion, a handful of extremely distant objects did not constitute a pattern that could overturn existing theories such as cold dark matter, which relied largely on statistical arguments.

However, the new superclusters were a more vexing problem. Still unseen, unknown and unaccounted for, the biggest of all, the Great Attractor lay hidden out there beyond Hydra-Centaurus, drawing Marc Davis's Cray and everything else in our neighborhood of the cosmos toward it in a great gravitational rush. Some theorists in the cold dark matter camp speculated that the secret attractor itself might very well be composed of some form of the strange dark stuff.[5]

Margaret Geller, whose work had played such a leading role in devastating many of the theoretical models about galactic formation, was rewarded in 1990 for her fine efforts with a five-year grant from the John D. and Catherine T. MacArthur Foundation. She thought most of the theoretical efforts were nonsense. She believed in the veracity of the data pouring into her Cambridge office from the Arizona telescopes. Geller had little use for the theories purporting to explain the observations. She, of course, did not have the answers herself. As far as she was concerned, the current theories were flawed because they ignored the data, the measurements, the real observations about the real universe. Theorists had simply stopped paying attention to what was actually going on out there.

"When I hear about this dark matter, it sounds like the ether," said Geller. "What is it? Where is it, this stuff that explains everything?"[6]

Here a Theory, There a Theory

Oddly enough, one of the beneficiaries of the devastating new observations was the Ostriker-Cowie mechanism for galactic formation. The explosion model had been lying around for years, considered by almost everybody a little theoretical backwash of minor interest. Owing to an eccentricity of Ostriker's, there had been an inherent suspicion of it all along.

"Jerry likes things that go bang," Carlos Frenk, the proponent of cold dark matter, explained.

In the late 1980s, with ever-greater galactic structures filling up the computer screens, theorists pulled out Ostriker-Cowie again, dusted it off and took another look. It went bang, all right, but it also offered a nongravitational explanation for some of the observed distribution of galaxies. An explosion model, some theorists believed, might offer some relief from a blind adherence to the strict rules of gravitational physics that were confounding many of the finest minds and most powerful Crays.

There was another nice aspect of the explosion perturbation model, as it was being called. It included a natural mechanism for separating out the mysterious dark matter from the ordinary, visible baryonic stuff of stars and galaxies. Ostriker and Cowie's explosions certainly were more than capable of shoving around the hydrogen and helium atoms that would soon be condensed into the visible stars of a galaxy.

But the explosion would have no effect at all on weakly interacting dark matter. This would remain in the original galaxy where it had begun, accounting for 90 percent or more of its mass for billions of years to come.

These were dangerous times for theories, though. By the end of the 1980s the cumulative evidence of bigger and bigger galactic structures was like an invincible heavyweight in his prime knocking each new contending theory out of the ring. Like most others, the explosion model lasted but a few rounds. Its fatal flaw was scale.

Just how big a structure could a chain-reaction explosion create? Upon closer examination, it looked like the upper limit was a cluster about only 20 million light years across, maybe a little more if the model's parameters were pushed to their theoretical limits.

Not one to believe in a strident loyalty to his own theories, Ostriker happily backed off. After all, Geller and Huchra had found structures up to half a billion light years across. Moreover, evidence existed that they might actually be larger. And that, Ostriker realized, was quite a gap to bridge. Cold dark matter computer models could not account for the galactic clusters or their apparent age. The Ostriker-Cowie mechanism lacked the punch. Hot dark matter in the form of the fast-moving neutrino apparently had failed.

Yet the great clusters _did_ exist out there, taunting the theorists with their presence. Where did the huge clusters come from?

Something was going on out there in the cosmos that was weirder than dark matter, did not rely on gravity to pull the clusters together and was more violent than Ostriker's explosions. By the early 1990s a growing coterie of cosmologists working in conjunction with particle physicists thought there must be something else. This came in the form of a new kind of theoretical phenomenon called the cosmic string.*

The origin of cosmic strings went back to the 1970s. In those years particle physicists were striving to create a theoretical union of the three subatomic forces: the electromagnetic force, the weak nuclear force, and the strong nuclear force. Particle theorists such as Sheldon Glashow and Steven Weinberg thought it was likely that the three forces had been united in one single force for a vanishingly small fragment of time after the big bang.

In 1974 Thomas Kibble, a theorist at Imperial College in London, became curious about the cosmological consequences of these early attempts at unifying the forces, the so-called grand unified theories, or GUTs. Kibble set about the work of categorizing the scars that appar-

---

*Cosmic strings should not be confused with superstrings. Superstrings are vibrating stringlike entities that make up the most fundamental particles, according to a recent and highly speculative theory.

ently would occur in space-time at the instant the three subatomic forces separated.

The scars, or defects, already had materialized as mathematical possibilities in the early attempts at grand unification. Many physicists thought, or at least hoped, that at the exact instant of the big bang, all the forces and particles were utterly indistinguishable from one another in a single state. Then in a sudden transition the forces broke apart and took on their individual identities.

According to the grand unified theories, this momentous event took place exactly 10^{-35} second after the big bang. Not coincidentally, this was precisely the same instant in early cosmic time that the inflationary expansion of Alan Guth, who had once been a particle physicist, had supposedly begun. Time and space, previously ablaze with saturated, primordial energy, suddenly cooled.

In the agony of the phase transition from hot to cold, the fabric of space-time may have been pulled apart along thin lines. To draw an imperfect but telling metaphor, the universe's swaddling clothes ripped apart as it suddenly ballooned from a size 0 to a size 10^{50}. Kibble thought that the rips in the structure of space and time might appear as long, exceedingly slender strands of nearly pure matter and energy. This was a startling idea. In theory it would mean that actual remnants of the big bang would endure within cosmic defects that would still exist today.

They would manifest themselves in a network of riplike defects throughout space and time, a sort of meandering cosmic museum containing relics of the conditions of the universe at the instant of its birth. Within the defects, the instant of phase transition would still prevail with all the particles and all the forces mingling together indistinguishably in a seething, primordial stew.

Kibble's work, though highly speculative, did succeed in drawing attention. Theorists found the concept attractive: remnants of the big bang itself still swinging through the universe in minuscule tubes of false vacuum. But what were the strings good for? None had, of course, been detected. Yet because of their compelling weirdness, cosmologists created a fairly large body of theoretical research into them without any particular idea as to their possible effects.

Alexander Vilenkin, a Tufts University cosmologist who admitted he had a habit of being drawn to the most outlandish ideas, believed that cosmic strings must have played a role in galaxy formation.[7] The strings, though trillions of times thinner than the diameter of a proton, would create an enormous gravitational field owing to their incredible density: Each inch would have a mass of 10^{16} tons. That was as great as an entire mountain range on Earth.

Vilenkin thought that when the universe was very young, small strings no more than 100 light years or so in length might have begun drawing in huge quantities of matter. Over time they would oscillate through space like colossal jumping ropes, forming loops and swirls. Where they intersected with one another, they would leave galactic seedlings. Larger loops would account for clusters of galaxies. Unusually long, unlooped strings would plow through the primordial universe leaving vast wakes that eventually were to become the great sheetlike galactic clusters.

All very well and good. But where were these ancient strings with such intense gravitational fields that a single one could wreak utter havoc on our planet if it came within even a few light years? David Bennet, a young theorist at the Fermilab National Laboratory in Batavia, Illinois, and François Bouchet of the University of California at Berkeley brought in heavy hitting computers to the rescue. They hoped that a potent Cray-2 could simulate the evolution of cosmic strings and dredge up at least a clue.

In their simulation, the computer screen lights up with a wavy, jumbled assortment of lines resembling an abstract expressionist painting. Gradually the computer strings are carried outward by the expanding universe, dispersing as they go. In the universe today, according to the Bennet and Bouchet models, the strings would have an average loop circumference of about a million light years.

Not to worry about a close encounter with one of these potentially devastating strings. The nearest would be more than a billion light years from the Milky Way. Only a very few of the longer, unlooped strings would still exist. Maybe fewer than half a dozen. Slithering like fishing worms in a can, these long strings slowly and continuously generate additional looped strings in order to maintain a steady supply.

In the mid-1980s cosmic strings were more wishful thinking than theoretical breakthrough. Then in 1985 Neil Turok, a handsome young English physicist with Clark Kent's horn-rimmed spectacles and firm jawline, momentarily got everyone's attention. A former student of Kibble's at Imperial College, Turok was well versed in the lore of strings.

Seeking some kind of direct evidence for them, Turok one day came across observational data relating to certain groups of galaxies known as Abell clusters. Studying them, he suddenly realized that the Abell galaxies appeared to be scattered throughout space in almost exactly the same distribution pattern computer models projected for cosmic strings.

It started to look like strings just might explain the structure of the universe. Suddenly everybody was interested.

Along with Andreas Albrecht of the Fermi National Accelerator Laboratory, Turok worked on model computer universes. In the computer projections great looping strings correlated very well with the observed distribution of galaxies. Moreover, they did so without relying on the biasing methods used in the cold dark matter computer universes. There were no fudge factors in the cosmic string universes, Turok observed. They were either right or they weren't.[8]

By the end of the decade, it was looking as if they weren't. The computer simulations showed that the strings probably were too small and too unstable to create galaxies. And unfortunately, there was still no direct evidence that strings existed.

In 1986 the theory's backers began trying to change that by exploiting an effect called gravitational lensing, first predicted by Einstein in 1936.* Einstein knew that a massive celestial object would have a gravitational field strong enough to deflect light, just as a glass lens is able to bend parallel light rays and focus them on a single point.

If a star were perfectly aligned with a massive object such as another star and the Earth, then light from the more distant star would be bent around the black hole by its gravitational field and appear on Earth as a bright ring. If the alignment were not perfect, observers on Earth would see a pair of distant stars. Einstein published his theory but concluded that astronomers would never be able to make use of this gravitational optometry, because in his opinion stars were too small and too numerous in our neighborhood of the universe.

In the late 1930s, the ever-prescient Fritz Zwicky, who was to set the stage for the dark matter perplexities, said he thought this probably would not be the case. He argued that *galaxies,* not stars, would be the most likely agents of the gravitational lensing effect. Galaxies were so plentiful that sooner or later astronomers were bound to find a pair, one near and one far, perfectly aligned with Earth.

The idea was relegated to obscurity for decades. It was not until 1979 when astronomers observed a double quasar (known formally as 0957 + 561 from its sky coordinates) in the constellation Ursa Major that the first gravitational lens was seen. Since then about twenty more candidate lens systems have been identified. But only about half a dozen have been reliably established by multiple observations.

Vilenkin, riding the first wave as usual, pointed out that a cosmic string could be a novel kind of gravitational lens, a powerful deflector of light from its immense gravitation. In fact, such a gravitational field would be so powerful that Bohdan Pkczynski of Princeton was led to predict that double images would be found that would be widely separated by large arcs of open sky. Such a gravitational mirage might appear as a series of parallel double images strung out on a line. In the mid-1980s several sightings were reported, but quickly shot down.[9]

The evidence was slimmer every day. Finally, the peripatetic Ostriker, having more or less abandoned his explosion theory, stepped in and tried to save the day. He and Edward Witten of the Institute for Advanced Study in Princeton suggested that the strings could be super-

---

*The first recorded mention of a possible gravitational lensing effect was made by Orest Chwolson in 1924, but his suggestion was vague and attracted little attention.

Cosmic Strings Cosmic strings, looping through the early universe on the left, begin attracting matter that eventually evolves into galaxies on the right.

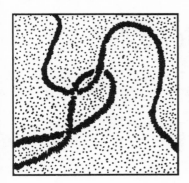 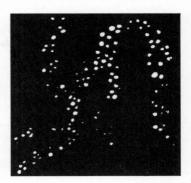

conductors, meaning that they could conduct an electric current with no resistance. If this had been the case in the early universe, they speculated, the strings would have emitted vast amounts of intense radiation that would have pushed and shoved the primordial gases into place for proper galaxy making.

The idea seemed merely an afterthought. As Witten observed, no theory required strings to be superconductors. It was just a possibility. In fact, it was beginning to seem that there was no particularly compelling reason at all for strings themselves to exist.

"The picture is not as nice as it once was," admitted Kibble.[10]

In the meantime, Turok jumped ship, volunteering to donate his computerized string codes to anyone who still happened to be interested. Now on the Princeton faculty, Turok worked for years on a new theory of space-time defects he believed might have better luck solving the twin problem of early homogeneity and late irregularity. Called textures, the latest theory in the cosmic sweepstakes was similar to string theory. Textures were knotty, local defects that Turok tried to show would perform better in computer models.

By now other theorists had become wary. Nobody was about to jump on anybody else's bandwagon until they had ascertained that it was going somewhere. Cold dark matter had looked promising. So had Ostriker's explosions at one time. And so had cosmic strings. All were in their own ways equally weird and thus compelling to theorists.

Yet none of the theories had been proved. Worse still, none even seemed remotely capable of being proved. If it wasn't superconducting cosmic strings or cold dark matter or explosions or textured knots, then what was it?

"It's likely to be something as bad, or worse," observed Ostriker mirthlessly.

CHAPTER NINE

Paradigms Lost

*No one was really accepted in Chicago until
he'd rubbed someone out. It was time for
Aristotle to get his.*
—Robert Pirsig

It's not even wrong.
—Wolfgang Pauli

What
was going on in cosmology? Thomas
Kuhn had explained it years before. Strangely, though,
it was not yet apparent to most cosmologists even by the early1990s.

In 1962 Kuhn revolutionized the history of science with a new vision of how science progressed. Before Kuhn almost everybody thought
that progress was achieved through the accumulation of detail. Each
new increment of knowledge was simply piled on top of the great body of
science. Isaac Newton had codified such an idea in a famous remark that
was sometimes thought of as Newton's other law: "If I have seen farther,
it is by standing on the shoulders of giants." [1]

The Trouble with Aristotle

As a young graduate student in theoretical physics at Harvard in
1947, Kuhn had been troubled by the absolute wrongness of early scientific ideas. One day he was asked if he would give a series of lectures on
the history of mechanics. He agreed, then immersed himself in the topic.
The trail led him back to the roots of the science of dynamics in fourth
century B.C. Greece.

Reading Aristotle's *Physica,* Kuhn was shocked at what he found.
There was no denying Aristotle's enormous wisdom, his pungent writing
and the originality of his thought as one of the creators of science. This
was what Kuhn had expected. However, he was astonished at something
else he had found. Aristotle had been terribly wrong in his thinking
about dynamics.

Kuhn was especially bothered by Aristotle's ideas about gravity. According to Aristotle's theory, the heavier an object was, the faster it would fall. He apparently had confused the effects of gravity's pull with the distance something could be thrown. The spell of Aristotle's theory had not been broken for two millennia. At last Galileo, the first legitimate modern scientist, decided to see for himself what actually happened to an object on the way down.

The more he thought about it, the more Kuhn was troubled by Aristotle's ideas. In every field other than physics Aristotle's observations had been acute and penetrating. In biology and political behavior, for instance, his ideas had been right on target. But when it came to motion, Aristotle's remarkable talents had failed him utterly.

"How could he have said so many apparently absurd things?" Kuhn wondered. "And above all, why had his views been taken so seriously for so long a time by so many of his successors? The more I read, the more puzzled I became. Aristotle could, of course, have been wrong—I had no doubt that he was—but was it conceivable that his errors had been so blatant?"[2] The answer eluded Kuhn. It was simply impossible that a man of Aristotle's intellectual astuteness had been so wrong about the nature of motion.

One hot summer day Kuhn suddenly realized he had been reading Aristotle in the wrong way. In an inspired moment Kuhn saw that Aristotle had not been mistaken after all. He simply had looked at the world differently. From Aristotle's perspective, his theories about nature were completely compatible with what he saw around him.

Once Kuhn adopted Aristotle's way of looking at the world, Aristotelian science suddenly became crystal clear. It was not misguided after all. It was just another way of looking at nature.

A Pair of New Spectacles

Now enlightened, Kuhn read Aristotle with renewed fervor and saw the Aristotelian world spread out before him in a new light. It was now a logical, consistent universe. The more Kuhn read the more he realized how impossible it was to draw any kind of analogy between the Aristotelian world and modern concepts about matter and motion.

"I did not become an Aristotelian physicist," recalled Kuhn. "But I had to some extent learned to think like one."

His new reading of Aristotle had revealed a global change in the way people viewed nature and applied language to it. Science had not progressed since the days of Aristotle simply by adding increments of data or the piecemeal correction of earlier mistakes.

Twenty years earlier, historians Edwin A. Burtt and Alexandre Koyré had looked at the scientific revolution of the seventeenth century and seen the same thing. Burtt and Koyré recognized that the revolution had involved an entirely new set of metaphysical assumptions; the revolution in thinking occurred as the result of a change in concept rather than through a series of new discoveries.

Kuhn was galvanized and soon turned to books on the gestalt school of psychology, which interpreted perceptual phenomena as organized wholes rather than aggregates of distinct parts. According to gestalt, a visual image or a body of knowledge could not be analyzed in terms of its constituents. The sum was greater than the parts. Analyzing its parts, Kuhn understood now, was exactly what he had tried to do when he first began reading Aristotle.

"While discovering history, I had discovered my first scientific revolution," he said.

Gestalt diagrams had pointed Kuhn in the right direction. If you look at the figure-and-ground illusion in one way, you might see a field of apples. If you flick your perspective a little, you can discern an array of Ts. You can look at it one way or the other. There is no middle ground. Kuhn believed that the maze of ink marks signifying nothing was the way a body of scientific knowledge looked to the outsider or to the student. Perfect examples were a topological map or the photograph of the subatomic events that occurred in an atomic accelerator, or atom smasher.

"Looking at a contour map, the student sees lines on paper, the cartographer a picture of a terrain. Looking at a bubble-chamber photograph, the student sees confused and broken lines, the physicist a record of familiar subnuclear events," Kuhn stated.

"Only after a number of such transformations of vision does the student become an inhabitant of the scientist's world, seeing what the scientist sees and responding as the scientist does," Kuhn observed.[3]

During the ensuing years Kuhn began realizing that the body of knowledge for any particular scientific community was like a shared gestalt illusion. Viewing a common world through the same hue of colored spectacles, scientists shared assumptions, presumptions and expectations about their work. A student could earn entrance into a scientific

Gestalt Illusion

community by putting on the special gestalt glasses worn by established members of the group.

How could a scientific community change its spectacles? Kuhn realized at once that a switch to new glasses would occur only through a revolution. One world view would be replaced by another. It was as simple as that.

Take the case of Ptolemaic astronomy, the system of compound circles carrying stars and planets around the Earth. No other ancient system had performed nearly as well. The system was especially admirable in its predictions for the changing positions of the stars. As the centuries passed, Ptolemy's system conformed less and less to the best observations of planetary positions and was especially disrupted by the precession of the equinoxes.

As discrepancies mounted, astronomers tried to eliminate them by tinkering with the system. Eventually it became apparent that a discrepancy fixed in one place was likely to turn up somewhere else. By the thirteenth century prescient astronomers began realizing that astronomy's complexity was now greater than its accuracy. Alfonso X (Alfonso the Wise), king of Spain from 1252 to 1284, a patron of philosophers and scientists, declared that if God had consulted him, he would have built a better universe.

By the early sixteenth century, astronomy had reached a state of crisis. Europe's best astronomers realized that the Ptolemaic system could no longer solve its own problems. Domenico da Novara, a colleague of Copernicus, said such a cumbersone theory could not possibly reflect nature. Copernicus called the astronomical system he had inherited a "monster."

Recognizing this, Copernicus was able to make the break in thinking that led to modern astronomy. He traded in his Ptolemaic glasses for a new pair, and suddenly saw an entirely different universe open up before him. None of the facts had changed, only his assembly of them.

His vision was sharper. But Copernicus's predictions of the positions of the stars and planets were no better than Ptolemy's. Kuhn was sure this was the case in most scientific revolutions: A new idea simply gave a new explanation for old facts. And this, Kuhn realized, was why it was so difficult for scientists to accept a revolutionary new theory. Often full acceptance would not be achieved until all the elder members of a scientific community had died.

Extraordinarily cautious, Kuhn did not get around to publishing his ideas until 1962 in a book called _The Structure of Scientific Revolutions_. By then he had hit on the word _paradigm_ to describe the world view of any specific scientific community. Simple and elegant, Kuhn's idea of revolutions as the means by which one paradigm was changed for another quickly caught on.

In a single stroke he had explained so much that had seemed so confusing, uprooting existing ideas about how science progressed. Indeed, for many, Kuhn had changed forever the way *all* historical development would be viewed.

Paradigms had now been unleashed on the world, and Kuhn's paradigm-anomaly-crisis-revolution-acceptance-paradigm cycle was soon being cited in fields far from the history of science. People were now seeing paradigms everywhere: in political science, in religion, in the federal government, in academic bureaucracies, in the war in Vietnam, in the changeover of automobile models, in the raising or lowering of hemlines.

Ironically, Kuhn himself had created such a powerful paradigm for viewing the development of all civilization that it no longer seemed possible to look at the world in any way other than through Kuhn's own special paradigmatic lenses. The idea itself had taken on the aspect of a grand truth, yet it drew detractors. Scientists divided into two camps: the avid Kuhnians and the anti-Kuhnians.

John Bahcall, senior astrophysicist at the Institute for Advanced Study and a leader on solar neutrinos, was a believer. As far as he was concerned, science had changed through a series of revolutions. Aristotelian gravitation had been overturned and replaced by Galileo and Newton, who in turn had been replaced by Einstein. He believed that eventually the current paradigm also would fall. Still, Bahcall realized, most scientists probably rejected Kuhn's ideas.

"I think it was Leibniz who described philosophy as the discipline in which you kick up a lot of dust and then complain that you can't see," said Bahcall. "That's an attitude which many scientists share." [4]

The late Richard Feynman, the great particle theorist and humanist, was against anything resembling a philosophical dust devil. One day in 1985 in his office at the California Institute of Technology, I asked his opinion of Kuhn's theory of scientific development through revolutions. He wasn't even mildly interested in the subject.

"Historians and philosophers seem to have an awful lot to say about the way they think science works. Or how it ought to work. Or how it could work. Or about this or that scientific method," Feynman said. "But it's almost always simplistic, and not very much related to what really goes on."

Other scientists heard Kuhn out, but many of them didn't like the message. They read his historical analysis as an attack on the validity of any kind of scientific truth. Indeed, the word *truth* had appeared in *Structure* only as a quotation from Francis Bacon. And Kuhn himself had stated that he didn't believe that there was such a thing as a final, grand scientific truth lurking out there to be found. The millennia-long cold war between science and teleology, it seemed, had at last been joined in Kuhn's slim book.

Science certainly had progressed from humble beginnings, Kuhn acknowledged. But it was not directed toward any final, grand scheme. This new perspective on history seemed an almost perfect one for this our most relativistic of all centuries.

Right or wrong, Kuhn's revolutions did seem to provide a means of looking at a modern world: an era of situational ethics and flexible moral standards; a time in which an important part of science was based on a set of theories under the general heading relativity or a nebulous uncertainty principle; an age that had abandoned earlier principles of absolute right or absolute wrong, absolute good or absolute evil, and, apparently now, even the possibility of an absolute truth.

A Crisis in Cosmology?

By the 1990s cosmology no longer presented a consistent picture of the universe. Was the big bang in serious trouble? Or did it just seem that way because of a few troublemakers? What about the inconsistencies that were showing up in the standard cosmological model and its inflationary corollaries? Only a few years earlier many fine scientists had been certain they were on the straight-and-narrow path toward a final truth. Presumably this truth would reveal itself shortly in the form of a set of facts and theories that would at last reveal the secret of the universe.

But serious anomalies remained. What about inconvenient observations such as the persistently uniform background radiation and the monumental galactic structures? Did they symbolize only a temporary breakdown? Or was the inevitable destination at the end of the road nothing more than an illusion, just another final scientific truth soon to be discarded to history?

In 1991 in the living room of his large brownstone residence on Boston's Beacon Hill Kuhn discussed the problems facing cosmology. He lived just a short distance across the Charles River from M.I.T. where he was professor of the history and philosophy of science.

"Look," he said. He used the word often. "I'm not very well versed in it. But I have heard there are some difficulties.

"But look, you shouldn't be too surprised. You have to remember that cosmology as a science is still very young. When I was in graduate school forty or fifty years ago, cosmology could barely be called a science at all.

"It's come a long way since then, mainly with the addition of quantum theory and a better understanding of relativity, which nobody knew much about then. Today better scientists with better tools are working in cosmology than was the case forty years ago."

Kuhn's appearance had changed little since his teaching days at

Princeton where I had first met him more than twenty years earlier. Then he had seemed a decade or more older than he was; now he looked younger than a man in his late sixties. He was stern, serious, unsmiling, cautious and gentlemanly; he was genial but not amiable, still every bit the scholar.

Kuhn was obviously burdened by the pain caused by the persistent misunderstanding of his ideas. During the intervening two decades he had been accused of claiming that scientists were irrational, that science was nothing more than power politics or mob psychology. Kuhn believed that a careful reading of *Structure* showed that none of this was true. In fact, he was actually pro-science.

"For Christ's sake," he said. "They didn't read the book."

What did he think was happening in cosmology? Where would he begin if he were looking into the possibility that a crisis had developed in cosmology in the 1980s and 1990s?

"Look," he said, thinking for a moment, "you need to find evidence that a sense of crisis has started filtering through the consciousness of the community. You might get a few clues from anecdotal evidence.

"Is there a breakdown in communication between groups of scientists? What is the sense of the degree of difficulties among the scientists? When did it begin?

"Look," he said one last time. "There's one other thing. I'd also try to find out whether a large number of new theories had suddenly started springing up."

The Kuhnian Cosmos

By the beginning of the 1990s every cosmologist seemed to have a pet theory, yet no one theory was compelling enough to draw in a preponderance of adherents. Moreover, one of Kuhn's signals for impending revolution already was at hand. There was already a sense of crisis in the astrophysics community.

As early as 1987 some cosmologists had begun conceding that something was wrong; that year, at an international meeting on dark matter it was apparent that astrophysicists were hard pressed to come up with a theory that gave a clear answer to the all the troubling questions raised by the new observations. Scott Tremaine of the Canadian Institute for Theoretical Astrophysics at the University of Toronto summed up the general feeling.

"The dominant impression which I carry away from this meeting is that extragalactic astronomy has reached a crisis," he observed.

"The nature, origin and distribution of the dark matter and its role in galaxy formation and dynamics are issues whose resolution is likely to determine the direction of studies in galactic structure and cosmology for decades to come." [5]

Tremaine traced the origin of the crisis back to 1974 when Ostriker, Peebles and Yahil had suggested that galaxies had gigantic dark halos containing most of their mass. This was the anomaly that had set the comfortable astrophysics community spinning.

Before that, Tremaine believed, astrophysicists had shared a common paradigm. This had consisted of the assumptions that Newton's laws were correct, that the mass of galaxies was mostly contained in visible stars, that the universe was more or less the same everywhere, that galaxy formation was unbiased, that the universe was expanding at a certain rate, and so forth. Once the dark matter problem became fully apparent, word of it spread quickly through the community. Eventually the most eminent scientists began concentrating on the problem.

"I am sure you will agree that this has happened," said Tremaine. "For many scientists, discouragement and pronounced professional insecurity set in, generated by the persistent failure of the puzzles of normal science to work out as they should."

More and more experimenters had begun devoting themselves to magnifying and making explicit the breakdown of the paradigm. This had occurred with the work of Vera Rubin and her collaborators; Geller and Huchra; the Seven Samurai; and others. By and by many new theories had been proposed.

"But their proponents are generally not very interested in talking to one another since they have abandoned the old paradigm which was what they once had in common," said Tremaine.

How long would a crisis in cosmology be expected to last? According to Kuhn, it would persist until a winning theory emerged from the group of contenders. He had estimated, based on his statistical work, that the time interval between the breakdown of the old paradigm and the enunciation of a new one was generally ten to twenty years. Tremaine thought that the current crisis had begun in 1974. In that case, there weren't too many years to wait for a resolution.

In fact, the crisis actually had started in the late 1980s, about 1987. That was the year most cosmologists were fully aware that they faced serious difficulties. If you believed Kuhn's figures, then the crisis probably would not be resolved until sometime early next century.

By 1990 matters had come to such a head that Jim Peebles decided he had to take matters into his own hands. He was just the man for the job.

Despite an outstanding reputation among his colleagues, Peebles had never received the media attention accorded other cosmological lights such as Hawking or Guth. This was fine as far as he was concerned. His job was to provoke and stimulate other physicists, not the general public. He was not even very interested in the literature of his own subject and was wary of large international physics symposia that drew the scientific press.

"If we met half as often, we wouldn't be so far behind," he once said.

He preferred small, informal meetings. Being an avid outdoorsman and skier, his favorite venue was the Aspen Center for Physics. At small gatherings there he liked to sit off to the side near the front in order to be best positioned for prodding and chiding the speaker, especially if he or she were one of his former students. Tall and rail-thin, Peebles would sprawl out, his legs and arms akimbo, his earnest and open face making him appear unusually benign.

However, this was merely a pleasant deception, for he was the master of the droll, sarcastic put-down, along with a humorless grin that spoke volumes. Peebles was the acknowledged theoretical maestro of such meetings, the cosmic grandmaster whose role was to usher new theoretical acts into the ring for a moment of glory perhaps or, more likely, to give them the hook.

Peebles was well suited for such a role. He had graduated near the top of his class from the University of Manitoba where he had switched from engineering to physics. He had gone on to Princeton for his graduate work in the heady days when it was the center of the universe for the fledgling science of cosmology.

There he fell into the orbit of Robert Dicke, a sort of Renaissance man of physics who was one of the few people ever to work successfully as both a theorist and an experimenter. A friend invited Peebles to attend small Friday evening seminars Dicke ran for a selected few. They would eat pizza and drink beer and discuss the big questions of the day.

After earning his Ph.D. he stayed on as a postdoctoral fellow in Dicke's group. This was a crucial period for a young scientist, the little window of opportunity when the course of an entire career could be set. It was a bittersweet time. He and Dicke and the others at Princeton had fallen in briefly with Wilson and Penzias and their accidental discovery of the microwave background at nearby Holmdel.

Peebles had done most of the theoretical work and, like Dicke, narrowly missed a Nobel Prize, partly through chance and partly through his own mistakes.* But at least it could be said that he had been present the day cosmology went legitimate.

After finishing his postdoc, Peebles single-handedly went about creating the branch of cosmology that concerned itself with the structure of the universe rather than with its fate or initial conditions. As far as he was concerned, such abstractions could be left up to theorists like his

---

*Peebles neglected to read about the earlier theoretical work of George Gamow and his colleague, and thus failed to acknowledge it in his paper. Possibly as a result of the ensuing flap over who was first with the correct prediction, no theorists were awarded any share in the prize.

Gamow was especially bitter and said at a conference years later, "If I lose a nickel and someone finds a nickel, I can't prove it's my nickel. Still, I lost a nickel just where they found one."

colleague at Princeton, John Archibald Wheeler, and Stephen Hawking and his band of followers at Cambridge.

Peebles wanted to know why there was matter and how it had formed itself into galaxies. In 1971 he codified his ideas in the seminal book *Physical Cosmology,* which for the first time injected detailed physics into the calculations that described the physical processes of the expanding universe. The book sent an entire generation of young theorists and observers out into the cosmos to ferret out the secret of the galaxies, the building blocks of the universe.

During a 1974 sabbatical a few blocks away at the Institute for Advanced Study, Peebles finished writing another book, this one with the ambitious title *The Large-Scale Structure of the Universe.* At that time there were only a few hints of the sheetlike galactic structures that would be found in such abundance a decade later, and Peebles in those days was skeptical that they existed at all. They were barely mentioned in the book, which nevertheless became a classic in the field.

By the beginning of the 1990s Peebles was directing new ideas on and off stage so quickly that he decided he had to find out for himself what was going on. He joined forces with Joseph Silk of Berkeley. The two had gotten to know each other during an earlier stint of Silk's at Princeton.

They decided to put together a side-by-side analysis of the merits of the leading theories for the large-scale structure of the universe. Which one was the best? Or was each one really as bad as the next? They used a simple statistical comparison and included five general theories for the origin of galaxies: cold dark matter, hot dark matter, cosmic strings, Ostriker's explosions and baryonic dark matter.*

Silk and Peebles analyzed the models in terms of their fit with 38 different kinds of observational data, including the cosmic background radiation, quasars, the big galactic structures, small-scale clustering and the internal structures of galaxies. They selected the various phenomena on the basis of how well established they were by observers, how much they were used by theorists and how much Peebles and Silk *thought* they should be used by theorists.

The result, as you might expect, showed no clear winner. Two theories, cold dark matter in an inflationary universe and baryonic dark matter in a low-density universe, did emerge slightly ahead of the pack. But even their correlations with the data were not very good.

"To enrich, enlighten, even amuse those of our colleagues who are

---

*Silk himself was a leading advocate of a theory of baryonic dark matter. He argued persuasively in his article, "Shedding Light on Baryonic Dark Matter" (*Science,* 251:537, Feb. 1, 1991), that compact stellar remnants such as neutron stars and white dwarfs would be excellent candidates for the missing matter. They were abundant, had the right mass and were well situated in the outer halos of galaxies.

trying to assess the merits of rival cosmogonies, we have begun a modest programme of setting up a cosmic book of odds," they wrote in their paper called a "A Cosmic Book of Phenomena," which was published in the British science journal *Nature* in July 1990.[6]

They went on to describe the theoretical situation as "dismal." On the other hand, they said, they were very impressed with the number of observational phenomena they had been able to use in the analysis. They thought that a good theory should be consistent over a wide range with observed data. Yet this was not the case for any of the theories.

"Therefore we would not give very high odds that any of these theories is a useful approximation of how galaxies were actually formed," they observed.

Predictably, the reaction from the cosmology community was not enthusiastic. Silk and Peebles were accused of the worst of all sins: not behaving like physicists.

"This isn't science," said Peebles's colleague at Princeton, Edwin Turner, an expert on gravitational lenses and an aficionado of rock music, baseball and Japanese poetry. He was brandishing "A Cosmic Book of Phenomena" in his office a few months after the seven-page paper had rocked the cosmology world.

"I'm not saying that means it's worthless. It is interesting. And it did cause a stir. But it's just not science. It's statistics or history or sociology. Something like that."

Turner and other critics, though, had missed the point. Peebles and Silk had momentarily traded in their own paradigm-viewing spectacles for Kuhn's, in order to see the bigger picture. And that picture was not very pretty.

Something *was* wrong. It looked like cosmology had entered Kuhn time, a period of crisis with no end in sight. There was no longer a leading theory to light the way. Perhaps none of the contenders had gone far enough—cold dark matter, cosmic strings or the others. Or maybe theorists had overlooked an elementary detail. Perhaps they had even missed something really big such as a second or third inflationary expansion or another big bang or, heaven forbid, the possibility that the universe was simply infinite, without beginning or end or boundary.

"Maybe something crazy is needed. None of the standard models for formation of galaxies and clusters of galaxies fits very well with all of the data," Peebles said. "We're getting a little desperate."

More than ever before, the data seemed to have been laid down by the hand of God, and theorists were having a difficult time reading the scripture. By the beginning of the 1990s it was fair to wonder if even the greatest cosmic hagiographa of all, the big bang, had become fair game.

This, of course, was not only in the hands of the astrophysicists but also in the hands of the particle physicists, whose domain was at the other end of the cosmic scale.

CHAPTER TEN

THE CRUCIBLE OF TIME

*There are moments where time
suddenly stands still and leaves
space for eternity.*
—FYODOR DOSTOEVSKY

Things
are not exactly as they seem in
the southwest corner of Switzerland. On the surface it
looks normal enough. In the shadow of the Jura Mountains and the Alps,
one of Europe's loveliest cities, Geneva, is strung out along the west end
of a lake of surpassing beauty. Mont Blanc rises sixty miles to the east,
its perpetually snow-crested summit glistening in the sun. The arbored
lake shore is lined with Mediterranean mansions, the Palais des Nations, international organizations such as the Red Cross and impeccable
public gardens. The overall ambience is so secure, rich and civilized that
the nonpowerful, the nonwealthy and the poorly mannered seem to have
been banished by edict.

One of Geneva's most unusual and least known attractions lies just
a few miles beyond the town line. Drive west from the heart of the city
on the Rue de la Servette. Soon you'll come to the little suburban village
of Meyrin just past the Cointrin international airport. Ahead is the
Franco-Swiss frontier.

On your left behind a chain-link fence is what looks like the campus
of a technical college. Fields and comfortable-looking farms and villages
are off on the right. But 100 meters beneath your feet is the biggest machine on earth, a particle accelerator operated by CERN.*

---

*CERN is a French acronym for European Council for Nuclear Research, which has
changed its official name to European Organization for Particle Physics in order to avoid
the stigma attached to the word *nuclear* in Europe.

In this particle accelerator energy and matter interchange with such intensity that a nuclear explosion is just the pop of a cap gun by comparison. Brought to life within the accelerator are forms of matter that were banished from the cosmos after their first and only appearance a trillionth of a second after scientific genesis supposedly brought natural law, matter and energy and the entire cosmos into existence all at once.

That is the scene if you are a believer in the big bang. If not, well, there are an awful lot of expensive positrons and electrons traveling very fast around an immense subterranean circular tube, constructed at the cost of billions of dollars and thousands upon thousands of worker-hours.

Accelerators like the enormous one at Geneva are the experimental laboratories of particle physicists. Here, physicists have peeled back the various layers of the atom right down to the minuscule entities that make up the nucleus. Many physicists believe that the nucleus guards the secret of the big bang. Yet, despite the immense size and cost of the machines, they have not been able to pry open the last door.

Ordeal by Fire

Time on the CERN machine is precious. Physicists from most of the European countries as well as from the United States, the Soviet Union and China compete to use it to pick and probe at the inner structure of the atom; this is a universe unto itself and as distant from daily life on Earth as a quasar at the edge of the universe. Accelerators were once called atom smashers, yet now the search is on for something much, much smaller and more energized than an atom.

In theory a device allowing physicists to travel back in time in the direction of the big bang, the CERN machine stretches through miles of subterranean passages, penetrating beneath the Jura Mountains and capable of producing temperatures in excess of 30,000 trillion degrees Celsius. Such a temperature, its designers hypothesize, is similar to conditions that occurred but a fragment of an instant after the primordial fireball.

One autumn morning in 1989 I went down for a look with Vince Hatton, a cheerful, red-haired Brit who was in charge of day-to-day operations. I was lucky; visitors are almost never allowed. The machine usually ran for weeks at a time and produced as a nasty little side effect intense and potentially lethal radiation. Right now the accelerator had been shut down for about forty-five minutes for a repair. We had time for a quick look.

We hopped aboard a small elevator car. Thirty floors down we stepped out. Hatton engaged an elaborate safety system with heavy steel

The Large Electron-Positron Collider More than 100 meters below the country-side, the accelerator sends particles slamming together in the hope of learning the universe's most closely kept secrets.

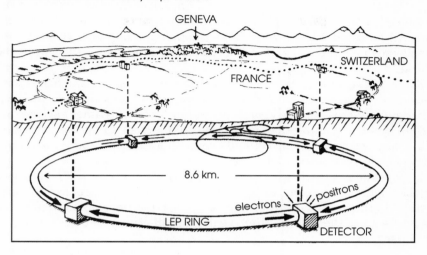

doors operated by electronic keys designed to keep humans out when the machine was running.

"The system is so good that it's suicide proof," said Hatton. He looked me over, then added without a trace of a smile, "in the event you're a person of such persuasion."

We walked into a bright, concrete-lined cavern. Squatting before us and looking for all the world like one of those plastic toys that can be twisted into different shapes—a warrior, a tank—was a gigantic piece of machinery: a particle detector. The accelerator's beam line stretched away from the detector through a gracefully curving tunnel, lovely and functional-looking in a Euro-tech kind of sturdiness and simplicity that conjured up a mental image of a snake of steel and concrete winding through a hole beneath the surface of another planet.

The accelerator was called LEP, the large electron-positron collider. It resided in a ring nearly 17 miles in circumference at an average depth of about 360 feet. It's almost impossible to make an overstatement about it, consisting as it does of 431,000 cubic yards of concrete and 60,000 tons of high-technology hardware.

Among the latter are almost 5,000 big electromagnets that propel and guide the electrons and positrons around the particle racecourse, 4,000 miles of electrical cables, nearly 200 computers and 4 monumental detectors, each weighing in at about 3,000 tons. Hundreds of scientists and engineers, all supposedly with the expertise to push the right button at the right time, were connected by a vastly complicated network of computers and electronic monitoring gadgetry to this technological maze.[1]

Unbent before the most profound and unyielding mysteries of the cosmos, Stephen Hawking was overwhelmed when he visited the collider. All he could muster was, "This reminds me of one of those James Bond movies, where some mad scientist is plotting to take over the world."[2]

Swinging around the accelerator's beam line, faster and faster with each circuit, particles reach velocities just below that of light, stateless little entities crossing the Franco-Swiss border 50,000 times each second before smashing into other particles coming from the other direction. Akin to thousands of cannon balls smashing into each other, the collisions are so violent that debris flies out in all directions.

That debris is what matters to the particle physicists. It is the focus of their search and contains, it is hoped, one of the secrets of the universe—in the form of newly created particles that have not existed in the universe since the instant after creation. If, of course, you are a believer in the big bang.

Down beneath the surface there was little sense of the wonder of the cosmos or the miracle of creation. The air smelled of oil and machinery. Hatton began explaining the detector's operation.

Klaxons sounded. The machine was about to be cranked up again. Time to go. We headed for an exit through one of the security doors so massive they looked like they belonged in front of a bank vault. It was jammed shut.

"Don't worry. They can't turn it on while we're in here," he said.

A moment passed, then another. Hatton actually looked a little nervous. He picked up a telephone to a control room. We waited.

A half mile away somebody finally pushed the right button. The door swung open and we were set free. Technology abhors a vacuum, and within minutes particles were again swinging along through the magnetized tube to the rhythms of the big bang.

In a control room above the particle detector, computer monitors blinked with colored graphs that charged subatomic conditions at two points along the ring. Computer operators adjusted the flow of electricity to the ring's magnets, steering the beams to ever-increasing rates of collisions.

"The trick to running an accelerator is to get the different parts together in synch," said Hatton. "Of course, the machine has a natural tendency to behave exactly the opposite."

He was scrambling to increase the number of collisions a hundredfold, no mean chore given that he needed to make dozens of rapid-fire adjustments among hundreds of thousands of parts.

"The more collisions, the easier it is to detect particle interactions," he said. His technicians were in a quiet frenzy. "We're in a race with every other big accelerator center to come up with major breakthroughs. Our funding, our very survival depends on it."

The scene was the same during a particle run at each of the world's ten major accelerator centers—in the United States, Europe, Japan and the Soviet Union.

There is no second place in physics, Nick Samios, director of the Brookhaven National Laboratory on Long Island, had told me once. "It's just like a football game. Nobody remembers who lost the game. And nobody remembers the second person to say $E = mc^2$."

Accelerator Pool

Running an accelerator was more or less like trying to play pocket billiards at night on a table as big as Texas with a cue ball that you can't see that you aim at another ball coming at you from the opposite direction at the speed of light.

The players in accelerator pool did have a slight advantage. They could shoot thousands of cue balls, and there were a huge number of balls coming the other way. Most missed each other. A few hit. The scoring would come from the detection of new subatomic particles created out of the energy transformed by the force of the collisions.

Atoms, though, are messy little objects, actually not objects at all. The nature of their constituent parts seems in large measure to depend on how we observe them. By our observation we alter their states. Can a child examine an ant without destroying it? In accelerator pool, the players, it seemed, were always behind the eight ball.

As elusive as they are, the parts of an atom have behaved in a manner that to a high degree is statistically predictable and thus suitable for scientific study. As the stakes of the game have gone higher, the players in accelerator pool have aimed bigger and bigger machines at smaller and smaller targets.

Thus the paradox of the accelerators: To see the smallest objects in the cosmos required the biggest machines of all. Accelerators have managed to pull back layer after layer of matter from molecules down to atoms. But atoms were produced when the universe was 10,000 years old and are far removed from big bang conditions imagined by cosmologists.

Continuing this artichoke-peeling procedure, the machines dug down into the atomic nucleus itself, pulling its protons and neutrons apart to reveal their inner structure: a tightly trussed framework built of quarks, the candidate of the moment for the smallest material entity.

The grand prize in accelerator pool would be confirmation of the big bang itself. There is one slight problem, though: no matter how big the machine and how advanced the technology, big bang energies are simply too staggering ever to be duplicated on Earth.

Master of the Game

This difficulty, of course, has not deterred accelerator builders in the least. Pushing on in the expectation and hope that revelation was just around the corner despite the cold, hard facts of scientific and technical reality, particle physicists developed two basic kinds of machines. One was a linear accelerator like the one operated by Stanford University near Palo Alto, California. Two miles long and as straight as the laser beam that was used to align it, the accelerator fires negatively charged electrons at atomic nuclei at 99.99 percent the speed of light.

Most of the rest—including the CERN machine and the biggest in the United States at Fermilab in Batavia, Illinois—fire their projectiles around a circle. Some use protons, which are heavier and generate more collisions, but it is easier to analyze collisions in the accelerators that use electrons.[3]

To record the collisions, particle physicists first built detectors called cloud chambers. Particles from collisions swept through water vapor like a truck through a drizzly mist, leaving a spray of droplet tracks that were photographed for analysis. The modern answer to the old cloud chamber, which was severely limited because of the large size of H_2O molecules, was the bubble chamber. David Glaser and some other physicists got the idea for it while watching beer foam.

When a particle passed through a bubble chamber, it collided with hydrogen nuclei, creating a spray of highly energized particles that scattered wildly. In their wake, the liquid begins to boil, and at that instant the cameras fire to record the particles' tracks for posterity.

It all seemed like a game of hide-and-seek in Alice's universe, a backward wonderland where cake was passed around before being sliced and Alice approached the Red Queen by walking away from her. Carlo Rubbia, CERN's director-general, cheerfully agreed when I suggested the analogy.

"It's actually even worse that," Rubbia said, laughing.

"It's like trying to tell the color of invisible jerseys worn by invisible football players in an invisible stadium by watching the movement of the ball.

"But, by God, when you find something new, it is an exhilarating moment."

Rubbia had had more than his share of these moments, which he thought of as fun. "Physics is fun," he said. "Forget about everything you've ever heard about it. Physics is fun."

Huge, charming and cunning, he was a man of legendary—and to his many critics and enemies, exasperating—energy. He was so big and wound-up, so animated and nervous and gesticulated so wildly, frequently waving his arms in great circles, that it was difficult to tell if he was charlatan or clown, genius or rogue, or, maybe, even a human pipeline to the mind of God.

Against the advice of many of the world's best particle physicists, including several Nobel Prize winners, Rubbia had bullied and badgered the CERN community of nations into building a huge proton-antiproton accelerator, LEP's predecessor at Geneva. His detractors claimed that the machine would be a waste of money for the simple reason that it would not work.

"These remarks were coming from very, very good people," said Rubbia. "I was scared stiff the beam wouldn't work." He kept quiet, though, knowing that public expression of his doubts would doom the accelerator. After the machine was built and found to be functioning perfectly, Rubbia took charge of it with another colleague, Simon van der Meer, who had conceived of a technically farfetched technique called stochastic cooling to gather and tend the herd of antiprotons with which Rubbia intended to keep the machine fed.*

Rubbia and van der Meer then geared up their new billion-dollar antiproton-gobbling toy to track down and eventually find a special class of particles called W's and Z's that would confirm a basic tenet of quantum theory: The force of electromagnetism and a seemingly disparate force, the weak nuclear force responsible for certain types of radioactive decay, had shared a common ancestor an instant after the big bang.[4]

For what marked the culmination of one of the century's greatest experimental chases, they were jointly awarded the 1983 Nobel Prize in physics. In the process, they also forced the biggest accelerator facility in the United States, Fermilab, into a scrambling game of catch-up while shifting the balance of power in high-energy physics from the United States to Europe for the first time since World War II.

Rubbia and van der Meer had created the model for the next generation of accelerators. Rubbia also set the standard for explaining how accelerators worked, too. He had a stock answer, invariably accompanied with great gusto and fierce Italian gesticulations.

"The basic principle is a little bit like smashing two cars together to find out how they work by seeing what falls out," he said. "When you smash two cars together on the highway, they are destroyed. In particle

---

*Stochastic cooling was a complicated technology that involved making the antiprotons move in a more coherent and orderly arrangement through the use of radio-frequency waves; thus many more antiprotons could be added to the beam line of a smaller accelerator nearby that was used to store the supply of antiprotons.

physics, when you smash two cars together, you get twenty or thirty new cars. Maybe even a semi-truck or two comes out."

The same principle has been used since 1932. One day that year a usually reticent young English physicist ran wildly through the gothic streets of Cambridge, accosting absolute strangers with the news.

"We have split the atom! We have split the atom!"

The man was John Cockroft. He and a colleague, an Irish physicist named Ernest Walton, had just built an extraordinary new machine that accelerated protons to high speeds and used them as projectiles to transmute an atomic nucleus, or break it apart into its constituent protons and neutrons.

The mechanism was the same as the one used in the primitive accelerator you probably have in your home. Inside a television picture tube, electricity heats a metal filament, boiling off negatively charged electrons and accelerating them through a positively charged wire grid. A magnet then steers them at the phosphorus-coated screen, which glows from the collisions. Thus arrive *The Simpsons.**

In many of the big high-energy labs the first step in accelerating subatomic particles up to today's velocities near the speed of light is a machine that is still called a Cockroft-Walton generator. The generators can extract protons (or electrons, in the case of electron-beam accelerators such as the one at Stanford) from hydrogen gas supplied in ordinary commercial bottles.

In 1978 Ernest Walton visited Fermilab. He was then seventy-eight years old and a Nobel laureate, having shared the 1951 prize with Sir John Cockroft. Walton was taken on a tour of the big accelerator, and then shown a Cockroft-Walton generator, a monstrous contraption covered with enormous metallic bubbles for discharging excess electrical energy.

It looked from top to bottom like it had been built as a prop for an old horror movie, the kind of machine a mad scientist would fire up in order to resuscitate the dead. As Walton gazed on, the device, as was its custom, emitted a tremendous bolt of lightning.

---

*The power of today's huge accelerators is measured in electron volts. The more of these a machine can produce, the more deeply it is able to penetrate the atomic nucleus. One electron volt (eV) is about the energy a single electron gains as it goes from the negative to the positive end of a flashlight battery. With a little more energy, an electron will be stripped from an atom.

It takes a few million electron volts (MeV) to begin breaking the nucleus apart into protons and neutrons, but a thousand times more still, in the billion eV (or GeV, for gigaelectron volt) range, to propel particles with enough force to shatter protons and begin creating forms of matter that do not normally exist on Earth. The biggest machines in the early 1990s were operating at the trillion electron volt (TeV) level.

"Ah," said Walton, rubbing his hands in delight, "the machine knows its master."[5]

Cutting the Uncuttable

The literal meaning of _atomos_ in ancient Greek was "uncuttable." In Miletus, the birthplace of natural philosophy, Leucippus seized on the word in the fifth century B.C. to designate the smallest indivisible piece of matter, enriching the lexicon and eternally damning science with the concept. His proposition was that if you cut an object such as a loaf of bread in half, then again and again and again until you could do it no longer, you would inevitably reach nature's tiniest bit of matter.

According to this first atomic theory, the atoms were all the same in composition, but could differ in shape. What was indivisible, or atomic, about them was that they could not be physically broken apart. Democritus, one of Leucippus's students, extended the idea to distinguish between things as they really were and as they seemed to be. He conjectured that the world around us really consisted just of atoms in motion, which we experienced in a number of ways. Primary qualities such as shape, size and matter would be determined by the atoms themselves. Secondary qualities such as sound, taste, and thought could be explained in terms of the primary qualities, but not in terms of the atoms.*

Whether the atomic theory of Leucippus and Democritus was based on observation or was just a philosophical shot in the dark that sent science on a difficult or even misguided quest is simply not known. Aside from the enduring hope of physicists that it must be so, why did a single fundamental building block have to exist for all of nature, as atomic theory suggested? Hopes that the universe actually held a single great truth have persisted for centuries, but they may have been nothing more than a projection of the human mind.

Bertrand Russell, the British philosopher who won the Nobel Prize for literature, believed that the original atomic theory was not a fortuitous discovery. Instead, it directly involved the logical structure of scientific explanation itself and was thus all but inevitable.

"What is it to give an account of something? It is to show how what occurs is a consequence of the changing configuration of things. Thus if we wish to explain a change in a material object, we must do this by

---

*In this view, even the human soul was made up of atoms, although of a special kind that were more refined than the others and distributed throughout the body. Later Epicurus and his followers interpreted this to mean that death meant disintegration and personal immortality was an impossibility.

reference to changing arrangements of hypothetical constituents that remain themselves unexplained," observed Russell.

"The explicative force of the atom remains intact as long as the atom is not itself under investigation. As soon as this happens the atom becomes the object of an empirical enquiry, and the explicative entities become sub-atomic particles, which in their turn remain unexplained."[6] And so on down and down and down.

Any atomic theory could thus be viewed as an inevitable consequence of human thought. Was the concept also a philosophical idea needed to satisfy humanity's eternal psychological yearning for revelation? Down through the ages, the mind seemed to proclaim, yes, yes, there *has* to be a piece of matter that is the smallest of all. There has to be a grand secret of the cosmos. Two and a half millennia later particle physicists and cosmologists and almost everybody else still seem equally enthralled with the notion that nature must still be holding back the answers.

Throughout Roman times the atomos paradigm rivaled Aristotle's material world of air, earth, fire and water, then fell into decline during the Middle Ages. By the seventeenth century, atomism was revived, especially in Isaac Newton's work on light, which he conceived of as being made of "corpuscles," or particles. In 1808 British chemist John Dalton put new energy into the concept by arguing that there must be a corresponding chemical element for each kind of atom, and that everything else was made from combinations of those atoms.

By the end of the century all ninety-two naturally occurring elements had been found. Now things were becoming complicated indeed, with many kinds of basic matter to be explained rather than just four. In 1896 Henri Becquerel, who had a highly unusual interest in things that glowed in the dark, discovered radioactivity, later elaborated on by Marie and Pierre Curie. Such an emission seemed to indicate that atoms were not solid objects after all.

A year later, J. J. Thomson, working at the Cavendish Laboratory in Cambridge, England, experimented with a cathode ray tube and discovered that atoms of rarefied gas could emit negatively charged particles (later called electrons) that were much smaller than the hydrogen atom.

The uncuttable had been cut. Enter the subatomic world and an entire new universe of ever greater degrees of complexity. No longer would nature present a simple face to humanity. Never again would something uncuttable remain indivisible, although each new generation of physicists seemed certain that the fundamental building block it had discovered was the last one, that one of the secrets of the universe was all but at hand.

In 1910 in a drab, brick-lined laboratory in Manchester, Ernest

Rutherford, a New Zealand physicist living in England, began bombarding gold foil with alpha particles (now known to be fast-moving helium nuclei). (A few years earlier Rutherford had worked with Thomson in Cambridge and discovered alpha, beta and gamma radiation.)* Big, energetic and intimidating, Rutherford had a habit of bellowing at his assistants if they weren't working fast enough. As he bullied the charged particles at the foil, he and the others in the lab were astonished to find that most of the particles shot straight through the foil, but that a very few—one in 8,000 or so—bounced back.

His experiments and some others conducted by Hans Geiger, a pioneer in radioactivity and atomic structure, led Rutherford to conclude that the gold atoms must be mostly empty space; the deflections meant there had to be something small and hard inside. With these experiments, Rutherford had found the atomic nucleus and set the stage for twentieth-century particle physics. The technology is better today, but the principle he developed—similar to shooting bullets at a shroud-covered object and studying the ricochet to deduce what is inside—is similar to that used in modern accelerators.

In 1911 Rutherford produced an entirely new model of the atom. Its positive charge and most of its mass were concentrated in the central nucleus, and electrons carrying the negative charge rotated around this nucleus. He later discovered that the positive charge of the nucleus was carried by particles 1,846 times heavier than the electrons although their charges were equal; he christened them protons.

The Rutherford atom was a perfectly balanced perpetual-motion machine, so minuscule that 10 million of them laid end to end would measure only half an inch. Still, its interior was mostly empty space, the nearest electron at such a distance from the nucleus that if the nucleus were the size of a tennis ball the nearest electron would be about a mile out. In effect, each atom was a miniature solar system, and it was this iconography so many us grew up with in the middle decades of the twentieth century.

However, a new kind of complexity was about to enter from another direction. During the 1890s scientists were unable to explain how a heated object such as a chunk of black iron could glow in different colors. A conservative member of Germany's cultural and ruling elite and stern-faced professor of theoretical physics at Berlin, Max Planck, recognized the seriousness of the problem.

Planck realized that this was a major challenge to the Newtonian view that everything in nature—from particles to planets—had a well-

---

*Rutherford's work led eventually to the release of atomic energy, but he never believed the atomic nucleus would provide a source of usable energy.

defined set of properties such as mass, radius, charge, velocity and location. In the deterministic Newtonian universe, the future of everything would be set forever by its past position, velocity and so forth, in combination with the laws of nature.

In 1900 Planck concluded that the frequency distribution of the radiation (that is, the different colors) could only be accounted for if its energy was exchanged, not in a continuous flow, but by individual packets, or quanta. Energy moved not like a river but like drops of rain. Almost at once he realized that his new quantum principle would devastate Newtonian concepts of nature.*

Crisis in Light

No final truth about the universe has lasted. This was true for physicists at the end of the nineteenth century who believed in the Newtonian vision of nature. It will also be true for cosmologists today who believe they are on the verge of a theory of everything.

Radiation was one chink in the Newtonian monolith. Another one already had appeared on another front. Newton's concepts of absolute time and absolute space were ancient ideas he had updated. Both time and space were pictured as existing independently of the rest of nature. Time was a flow like a river, and space was the changeless, unyielding, fixed matrix for the universe.[7]

Newton's longtime nemesis, Baron Gottfried Wilhelm von Leibniz, was an inspired mathematician who was also official historian for the House of Hanover. Leibniz, with one or two others in the late seventeenth century, had questioned the idea of absolute positions and motion. They were *almost* able to show mathematically that such concepts were not really required for Newtonian physics. An encyclopedic genius who unfortunately published few complete works, Leibniz even went so far as to hint at the great mathematical appeal of a relativistic view of time and space, but he did not follow through with the idea.

The roots of the crisis in the nineteenth-century physics paradigm had appeared as early as 1820 with the general acceptance of the idea that light traveled in waves. If space and time were absolutes, then Newton's laws seemed to require that light waves needed a medium through which to propagate. There could be no waves on the sea without an ocean of water to carry them.

---

*Developed in conjunction with his radiation law, Planck's constant was equal to the energy of a quantum of electromagnetic radiation (one photon) divided by the frequency of the radiation: h (the constant) $= 6.626 \times 10^{-34}$ joule second. A basic element of quantum theory, it came to be regarded as a universal constant of nature, as fundamental as Einstein's c, the speed of light.

Scientists believed then that the medium for light was stationary ether anchored in a Newtonian matrix of absolute time and space. Light would propagate through this ether like waves of sound through the air. Aristotle had first used the word to describe a hypothetical fifth form of matter, a luminous material of which he speculated the stars and planets were made.

Newton and other seventeenth-century scientists had postulated the more modern incarnation of the theory. Comprehensive and easy to understand, the idea was so logical that, with advancing technology in their hands, scientists rushed to the "luminiferous aether." This was no easy feat since it supposedly was a miraculous form of matter that was utterly transparent and frictionless. However, Newton and others believed that as the Earth traveled through it at 18.6 miles per second on its trip around the sun, the planet's motion would stir up an "aether wind."

If light waves were carried by the ether, then the ether must offer at least _some_ resistance to them. This then would mean that there would be a difference in the speed of a beam of light when traveling across the ether wind or against it; this came to be called the famous ether drift.

Unfortunately, no drift was observed, even with expensive and specialized equipment built for the search. This forced leading mid-nineteenth century theorists on the nature of light—Augustin Jean Fresnel, Sir George Gabriel Stokes and others—to devise a series of highly speculative new hypotheses to explain the failures, much the way cosmologists today are scrambling to construct theories of galactic formation based on an invisible dark matter.

In order to detect the ether, the velocity of light had to be known. Galileo had first tried to measure the speed of light with an experiment in which he stationed two assistants on distant hills and had them flash lanterns at one another so he could time the interval. Not having access to a clock, he tried to use his pulse to time a period of about one hundred thousandth of a second.

Although measurements had improved substantially during the eighteenth and early nineteenth centuries, they were still not exact enough to detect the ether. In 1877 Albert A. Michelson, a young physics instructor at the U.S. Naval Academy in Annapolis, went to work to determine the exact speed of light. The son of a Polish immigrant dry-goods merchant, Michelson had graduated first in his class from the academy four years earlier and was uncommonly bright and a whiz at optics.

His experiment cost only a few dollars. It consisted of a lamp, a condensing lens and two mirrors 500 feet apart. One mirror was fixed, while the other rotated on a vertical spindle at 130 revolutions per second. Light focused on the rotating mirror was reflected through the lens onto the fixed mirror, then back again in flashes. The amount of displacement could be measured as an angle, which let Michelson calculate the speed of the light as 186,508 miles per second, a remarkably accurate figure.[8]

Michelson next took up the ether problem. He built an extremely sensitive optical interferometer, which worked by timing the mirrored return of two simultaneous beams of light. One would be fired at right angles to the ether wind, and the other directly into it. The slightest difference in times should be a way to measure the interference effect.

Edward W. Morley, a renowned chemist, joined forces with Michelson. Beginning in 1887, they tested for differences in the speed of light into the ether breeze and across it at all angles. The results were always the same: nothing. Yet, like the idea of the big bang today, the notion of the ether was so much a part of the conventional scientific wisdom of the day that they assumed the problem had to be with their equipment. They tested and retested the apparatus.

They worked on. The strain of constant failure eventually caused the obsessive Michelson to have a nervous breakdown. At last they were convinced. There was no ether wind. By extrapolation, there was no ether either.[9]

This was startling news. It came from two of America's finest and best-known scientists. Although no one doubted their results, the meaning of the results was not clear. Michelson and Morley had revealed a startling anomaly in the Newtonian paradigm similar to the inconsistencies that had begun plaguing cosmology one hundred years later.*

The resulting crisis, initiated by the Michelson-Morley experiment in 1887, was not resolved until Einstein's special theory of relativity emerged in 1905. This opened the way for the convergence, later in the century, of the atom with astronomy in the new science of cosmology.

Ironically, in a reverberation of history almost exactly a century later, cosmologists had begun facing their own first crisis.

---

*In 1907 Michelson became the first American to win the Nobel Prize for his "exactness of measurements." Later Einstein paid tribute to him: "Through your marvelous experimental work [you] paved the way for the development of the theory of relativity. Without your work this theory would be scarcely more than an interesting speculation."

CHAPTER ELEVEN

THE
QUARK
AND THE
COSMOS

Where the telescope ends,
the microscope begins.
Which of the two has
the grander view?
—Victor Hugo

Rain
fell in the morning. By afternoon
the sky was deep blue, the air as light and frothy as
champagne. The steep-walled valleys of the Elk Mountains in central
Colorado were alive with rushing water, the meadows awash in wild
flowers. On the west side of the old mining town of Aspen snatches of
Beethoven's Sixth Symphony wisped out from beneath a large canopy
where young musicians from the summer music school rehearsed for the
evening's concert.

Across a meadow a blackboard had been set up amid a grove of
aspens. Scattered among the trees was a collection of square wooden
buildings so ordinary-looking it was hard to believe they had passed the
Victorian resort's strict building codes. About a dozen men and one or two
women mostly in jeans, flannel shirts and hiking boots or shorts and
sneakers sat in folding bridge chairs. A harried-looking young man furiously scribbled equations on the blackboard, then just as frantically
erased them into oblivion.

The source of his irritation stood off to one side: a short, compact,
older man with curly steel-gray hair and glasses, he was dressed a notch
or two better than the rest in blue slacks and striped golf shirt. Issuing
a steady stream of brusque instructions and interrupting at will, he said
at last, "No, no, no. You've got it all wrong," and, grabbing the chalk and
eraser himself, took over the blackboard.

MGM

The place was the Aspen Center for Physics, the summer camp for the stars of cosmology and particle physics who came to hike and climb and go mountain biking, eat some of the best food in the Rocky Mountains, raft on white-water rivers, read the scientific papers they hadn't found the time for during the academic year and just relax. Oh, and they would also do physics, usually in the morning or late afternoon at their unpretentious little encampment in the fashionable west end. This left the middle of the day free for their mountain activities.

The best of the "phizzies," an endearment coined by members of the summer staff, had been here: Hawking, Peebles, Ostriker and the rest from the cosmology side; and Steven Weinberg, Sheldon Glashow, and everybody else from the particle side who could find the time and the money. One who had both was Murray Gell-Mann, the provocateur at the blackboard and tacit headmaster of the particle theorists in summer residence at Aspen.

MGM, as he was known around town, owned a bright blue, unrenovated Victorian house within walking distance of the physics center. It reputedly had been paid for with his earnings from the Nobel Prize that he had won at the tender age of forty. Very, very smart, extremely brash, often brusque, he was not above bullying a colleague whose argument he felt was proceeding too slowly for his own rapid-fire mental processes; nor would he hesitate to pronounce a redundant or boring question "foolish" or "totally trivial," if he chose to acknowledge it at all.

Or worse. Once on the grounds of the physics center a pretentious art director from a national geographic magazine pompously tried to explain the atom's interior to MGM, then asked his opinion of a large rendering of the atomic nucleus to be used in the publication. Gell-Mann held up the painting, dramatically and disdainfully turning it this way and that.

"What is this?" he asked. "I can't make heads nor tails of it." He then dropped the expensive painting face down on the wet grass and, placing his left foot on it as if he had not noticed, began grinding his heel into it.

The youngest son of Austrian immigrants, Gell-Mann was fifteen when he entered Yale, had earned a Ph.D. in physics from M.I.T. before his twenty-first birthday, and, following a brief tenure at the Institute for Advanced Study, was named professor of physics at Caltech when he was twenty-six. In school he had found physics boring, and, later in life, apparently considered being one of the world's great particle theorists little more than an incidental aspect of his life.

His many talents and interests were legendary; he had expert knowledge about subjects as diverse as mushrooms, geology and Afghan carpets, arms control and etymology. He was able to speak eight languages including Swahili and Mandarin Chinese, and he was familiar with at least that many more, including ancient Flemish, though he conceded he was entirely comfortable only in French.

It was evident that some unusual synaptic activity was at work, a fact that MGM himself would happily concede. If you were to accuse him of being the world's smartest man, he would not deny it; in fact, he seemed to encourage such remarks, an attitude some of his colleagues found somewhat galling.

The office of the late Richard Feynman, another candidate for world's smartest man, was just down the hall from Gell-Mann's at Caltech; they shared the same assistant, the efficient and delightful Helen Tuck. Once Feynman, who had had his own share of breakthroughs in particle theory and his own Nobel, was asked, "What are the most important ideas in particle theories?"

"Those of Murray Gell-Mann," quipped the perpetually irreverent Feynman.

Gell-Mann, of course, was eminently entitled to crow about his accomplishments. During a period of prodigious output in the early 1960s—not unlike that of Albert Einstein in the years up to World War I—Gell-Mann had brought order to the highly confused house of particle physics, then suffering under its own extended Kuhnian crisis. Almost single-handedly, MGM had bullied the particle physics community into a new vision of the subnuclear world, into the modern age and into what seemed to be an inevitable marriage with the cosmos.

The Particle Bestiary

The roots of the crisis in particle physics went deep. In 1905, the year that he developed his special relativity theory equating energy and mass, Albert Einstein proposed that light itself was quantized, or particlelike. This explained the mystery of the 1890s, how electrons were emitted when light hit certain metals, such as a hot block of iron.

Among the most fruitful periods in the history of science, the years following the turn of the century were a golden era, not yet duplicated, which brought about the great revolutions of the twentieth century. In 1913 Niels Bohr, a stout, young Danish physicist with a huge leonine head who had worked with Rutherford in Manchester, published a stunning series of papers in which he laid the groundwork for the atomic age.[1]

Relying on Einstein's special relativity, Bohr proposed that electrons behaved in a quantum fashion based upon the principle that had been described by Max Planck. The electrons would remain in fixed orbits in Bohr's theory, then move from one orbit to another in discrete quantum leaps when they emitted or absorbed energy. These orbits lay in separate and distinct shells around the nucleus. Bohr initially likened the interior of his atom to a minuscule solar system.

During the ensuing years Bohr gradually came to realize that this was not the case at all: Gravitation had no role inside the atom, which was governed by an entirely different set of natural laws: quantum mechanics. This recognition by the ever-questioning, always open-minded, occasionally absentminded Dane (who once removed a pipe from his mouth and couldn't find it until his coat pocket burst into flames), led to a decades-long debate with Einstein over the true nature of the universe and humankind's ability to understand it.

Fierce, compassionate and good-humored all at once, the argument between Bohr and Einstein, who had enormous respect and affection for each other, centered on the natural laws that governed the universe. Which laws predominated, the quantum laws of the atom or those of general relativity, which described gravitation and the universe at large-scale? Although most physicists at the time thought that Bohr had the better argument, he and Einstein made virtually no progress toward breaking the deadlock. In fact, by the 1990s the two sets of natural laws seemed every bit as incompatible as they had at the beginning of the century.

In 1920 Bohr founded the Institute for Theoretical Physics at the University of Copenhagen; throughout his life he was the institute's only director. Forty or fifty young graduate physicists from a dozen or more nations were usually enrolled. During the 1920s the institute was considered the "physicists' spiritual capital" because of the enlightened, questioning atmosphere Bohr had created.

"The unique and exciting feature of Copenhagen was in the climate of opinion, the assessment of ideas and the stimulus Bohr gave," said John A. Wheeler, one of Bohr's protégés and greatest successes. "I know of nothing with which to compare it except the school of Plato." [2]

Atomic physics had its great synthesis during the 1920s, a heroic period when breakneck progress seemed to be taking place every day. Although not the doing of any single individual, Bohr's deeply creative, subtle and critical spirit transcended the enterprise, while guiding, restraining and deepening it.

As quantum theory, the set of mathematical equations describing nature on the atomic scale, evolved, the very nature of reality seemed to become increasing elusive. Wolfgang Pauli, a big, brash Austrian who had a habit of offending his colleagues with the rapid put-down, had an unusual flair for mathematical articulation. He had an enormous ego and suffered from ubiquitous-knowledge syndrome, once lamenting that there were no more physics problems for him to solve because he already knew so much.*

---

*Pauli said of a colleague he considered an upstart, "so young and already so unknown." Once attending a seminar given by Einstein, Pauli, then a graduate student, stood up and declared, "You know, what Professor Einstein says is not so stupid."

In 1925 Pauli proposed that no particles of the class containing protons, electrons or neutrons could occupy the same quantum state at the same time. Thus, according to the Pauli exclusion principle, as it came to be called, no two electrons could occupy the same orbit, a theory necessary for understanding the chemical bonds between atoms. The principle also declared that there was a limit on the number of protons and neutrons that a stable nucleus could contain.

A crucial year in the young life of quantum mechanics was 1927, when a young German physicist named Werner Heisenberg formulated the uncertainty principle. He proposed that we could know either where a subatomic particle was at any given moment or we could know where it was going in terms of its momentum. We could not know both. Moreover, according to Heisenberg, the more precisely we knew the momentum of a particle, the less precisely we would be able to know its position. Heisenberg demonstrated the remarkable proposition by means of a penetrating mathematical analysis that showed that the limitations of the observer did not result from limitations in experimental technique now or at some imagined future time when those technical restrictions might very well be lifted.[3]

Rather, the limitations were imposed by the very nature of the subatomic world itself. It was *impossible* to measure simultaneously both the precise momentum and the precise position of a subatomic particle, not just technically difficult. We were, Heisenberg had showed, fundamentally limited in our ability to gather knowledge about the subatomic world.

The classical Newtonian version of the laws of physics had implied that knowledge of the exact position and momentum of everything in the universe—from galaxies all the way down to the smallest of particles in the subatomic world—should be theoretically, if not actually, attainable. Heisenberg's 1927 principle was the final break with the Newtonian universe. Nature on the atomic-scale, it appeared after all, was not determined by mechanistic laws.

Erwin Schrödinger, who worked with Einstein at the University of Berlin during the 1920s and got along quite well with him, constructed a set of equations that replaced Newtonian physics at the quantum level.[4] The solutions to these equations were mathematical functions that described only probabilities, not actualities. Schrödinger equations suggested that electrons were waves. Another friend of Einstein's, Max Born, theorized that the wave idea was useful merely for describing the probability for an electron's location. The quantum world was one where reality and illusion merged, a place in which observation not only affected reality but in a sense created it. Physicists themselves could choose to measure light as particles *or* as waves.

Beginning in 1927 Paul A. M. Dirac, a young Cambridge University theorist, began working on another front to attempt to create a mathematical union of Einstein's special theory of relativity, equating mass and energy, with quantum mechanics. The unexpected end product, antimatter, was as outlandish as quantum mechanics itself.[5]

While Pauli was brash and talkative, Dirac had withdrawn into mathematics and an utterly empirical view of the world. Bohr once said, "Of all physicists, Dirac has the purest soul." Dirac had little interest in things, only mathematical abstractions.

"A great deal of my work is just playing with equations," he once said, "just looking for beautiful mathematical relations which maybe don't have any physical meaning at all. Sometimes they do."

Stories about his legendary taciturnity circulated regularly through the physics community. Once a journalist attempting to interview him asked, "Professor, you have quite a few letters in front of your name. Do they stand for anything in particular?"

"No," said Dirac.

"You mean I can write my own ticket?" asked the journalist.

"Yes," said Dirac.

"Will it be all right if I say that P. A. M. stands for Poincaré Aloysius Mussolini?"

Dirac: "Yes."

In the 1920s theorists generally believed that there were just two elementary particles in nature: protons and electrons, each bearing the opposite charge. Before Dirac, quantum mechanics and special relativity were thought of as separate natural realms, with slowly moving electrons in atoms described by quantum equations, while special relativity applied to particles near the speed of light. Dirac brilliantly combined the two in a single set of equations that governed the behavior of electrons.[6]

The masterful equations made a number of predictions. One, already anticipated, mandated that electrons had spin. Another, totally unexpected, was that there existed a new particle: an electron with a positive charge. Dirac believed that the particle must be a proton, although he soon realized it was not: It had the same mass as an electron, while the mass of a proton was about 1,800 times greater. He called the new particle an "anti-electron."

Carl Anderson, a Caltech physicist, confirmed Dirac's prediction four years later, when he found the positive electron, or positron, in a cloud chamber. When a positron and an electron encountered one another in an accelerator, Anderson discovered, the two simply vanished in a mutual annihilation, the only trace a burst of X rays. Following Anderson's discovery, which had the beneficial side effect of winning both him and Dirac Nobel Prizes within a few years, bigger accelerators confirmed beyond a doubt that for every particle there was an antiparticle.

The absolute symmetry between normal matter and the new antimatter meant that if all the particles in any closed system were to become antiparticles, a distant observer could not tell the difference. All the laws of physics would govern either system. Physicists amused themselves with the idea for years, speculating endlessly about encounters between Earth and an anti-Earth or the universe and an anti-universe.*

But where *was* all the antimatter? According to Dirac's equation, later confirmed by thousands of accelerator experiments, the universe should consist of equal amounts of matter and antimatter. Yet there was no sign at all of antimatter on Earth, and, from studies of cosmic rays, which bore a few antiprotons from deep intergalactic space, there appeared to be no more than 1 part per 10 billion throughout the universe.

Something was wrong. Andrei Sakharov and other theorists advanced an explanation: During the first 10^{-6} second after the big bang, the universe may have contained a slight excess of matter over antimatter, variously estimated as about 1 part in 10 billion. Within the next fragmentary part of a second, matter and antimatter collided in a colossal annihilation, leaving behind only radiation and the surplus matter. What survived was the universe of today, its stars, its black holes, its galaxies, its single known solar system, its one known life-giving planet all consisting of this residual stuff. This was speculation, at best, and the conundrum persisted into the 1990s. If the matter-antimatter symmetry was so deeply apparent in theory and in accelerators, why did the universe itself still seem to consist almost entirely of matter rather than anti-matter?[7]

With quantum mechanics in place, the atomic age blossomed during the decade and a half after World War II, a glorious era for physicists who were becoming increasingly confident they could solve any problem set before them.[8] In the mid-1950s, Edward Teller, who had led the way in pioneering the United States's first hydrogen bomb, flatly predicted that unlimited energy from nuclear fusion was but five years away. The end was in sight for particle physics, with a full understanding of the atom, and therefore the entire universe, all but at hand.

Why the lack of skepticism? Perhaps the allure of the great cosmic secret, like a phantom light on a distant shore beckoning a doomed sailor, was simply too great. It was, of course, not to be. In a situation similar to that which had developed in cosmology by the 1990s, advancing technology, in the form of larger accelerators, soon made a mockery of every theory purporting to explain the particle world.

By the late 1950s big new accelerators in Europe and the United States were turning up scores of new particles. These were mostly in a

---

*In 1956 Harold Furth, a young physicist at the University of California at Berkeley, published a poem called "The Perils of Modern Living" in *The New Yorker*. Furth imagined a meeting between Dr. Edward Teller, so-called father of the hydrogen bomb, and Dr. Edward Anti-Teller. "Their right hands clasped," wrote Furth, "and the rest was gamma rays."

class called hadrons (from the Greek word for strong, because they were associated with the strongest of the three forces that governed the atomic world, the strong nuclear force).

By the early 1960s nearly 100 new particles had been found. They fit no theoretical pattern at all. Enrico Fermi, one of the greats of the day who was as adept at theory as he was at experimentation, complained vocally that had he known how many new particles were to begin cluttering up physics, he would have taken up zoology instead. Zoology was far better suited to classification, and order was desperately needed in the particle zoo.[9]

The problems in particle physics had reached crisis proportions.

The Eightfold Realm

Enter MGM. The early 1960s were heady years for particle physicists at Caltech. Feynman was there, developing the quantum field equations and Feynman diagrams that would lead to his Nobel Prize in 1965. Sheldon Glashow was a postdoctoral fellow, along with several graduate students who later would make names for themselves: Sidney Coleman, Bill Wagner, Ken Wilson, and an intense, swarthy fellow with an enormous mane of black hair named George Zweig.

Gell-Mann, especially, was feeling his oats during those years. Relying on his great energy and prodigious intellect, Gell-Mann worked at a furious pace to determine the underlying relationship of the newly discovered particles. In top form, his output during 1961 and 1962 was phenomenal, with a new paper appearing almost every two months.[10] Looking for a new tool, Gell-Mann turned to a branch of mathematics almost never before used in physics, abstract algebra. With this tool, he soon realized most of the new particles could be organized into groups of eight (octets) or, occasionally, by means of another clever patterning, into families of ten (decuplets).

The family groups of particles seemed reminiscent of the periodic table of the elements used by chemists. Gell-Mann named his organizational synthesis the eightfold way, an expression he borrowed from Buddhism.* The eightfold way is a spiritual progression for attaining enlightenment, devised by the Buddha some twenty-six centuries earlier.[11]

Gell-Mann's theory predicted the existence of an exotic new particle that he named the omega minus (Ω^-). Two years later, after an extended search, Nicholas Samios, a young experimentalist at the Brookhaven accelerator lab on Long Island, found the new particle in a bubble chamber,

---

*The symmetry was found independently at about the same time by an Israeli theorist working in England, Yuval Ne'eman, who did not apply such an imaginative name to his discovery.

providing confirmation that Gell-Mann's organization was correct. Still, nobody, including Gell-Mann, was sure what it meant. Something still seemed to be missing. Gell-Mann was at Columbia University giving a series of lectures while working on the problem. In response to a question at one of the lectures, he conceded that all the new hadron particle groupings could be simply explained if it were assumed that there were three more fundamental particles.

But, said Gell-Mann, the new particles would have to have "quirky properties." For one thing, they would have to carry fractional electrical charges—plus or minus one-third or two-thirds of the electron's charge. For another, the particles would have to have one-third of the mass and other properties of a proton or neutron. As he finished up the talks in New York, Gell-Mann began recognizing that this triplet idea was not such a crazy one, after all. It did seem to hold some promise.

In the meantime, George Zweig, who had been a graduate student at Caltech during the years when Gell-Mann was creating the eightfold way, had arrived at the same conclusion during an extended visit to CERN in Geneva. Encountering a great deal of difficulty in getting his CERN report published in the form that he wanted, Zweig finally gave up trying.[12] He did, though, have a name for his new particles: aces, from the expression, "dealer's choice—aces are wild."

Back in California, Gell-Mann had little difficulty publishing his new triplet theory. A master of nomenclature as a result of his literary and linguistic expertise, Gell-Mann had come up with a far more dramatic name for his new particles: quarks.[13]

"Quark means curd in German," Gell-Mann said over coffee with me one Sunday morning years later in the small, cluttered parlor of his Aspen home. "It was also slang in German for nonsense. I had liked the sound at once.

"Later I was reading *Finnegans Wake* by Joyce, and came upon the line, 'Three quarks for Muster Mark.' It was a drunken barman talking, and he must have meant quarts.

"In any case, there were three of them, and there were three fundamental particles in my theory. I knew the name was right."

The name and the theory stuck, and Zweig's aces were quickly forgotten.* Suddenly, the quark synthesis had reintroduced simplicity to nature. Almost everybody was delighted.

---

*The effect of Gell-Mann's nomenclature on the Swedish Academy was not clear when five years later he was awarded a rare solo Nobel Prize; in any event, he had laid the groundwork for quark theory in the eightfold way. When I met George Zweig in the early 1980s at the Los Alamos National Laboratory in New Mexico, he introduced himself rather sadly, I thought, as "the man who invented quarks." He conceded that the name "aces" had never caught on (perhaps because there are four, not three, of them in a deck of cards). In fact, he said, reaction both to his expression aces and to him had been "not benign."

Gell-Mann's second revolution, one of nomenclature, set another trend in initiating the era of whimsy. In order of ascending mass, the three quarks then postulated were called the up, the down, and the strange, labels having absolutely nothing to do with their relative vertical positioning or their eccentric nature.

The names distinguished the new particles according to properties of spin, mass and electrical charge, which was invariably plus or minus one-third or two-thirds that of an electron. The two older residents of the atomic nucleus, the proton and the neutron, could now be explained as combinations of quarks bonded together according to another property designated as color, which enabled them to join and form new particles. Other properties, including charge, determined a quark's flavor, that is, whether it was up, down or strange.

With an understanding of quarks, then, a proton could now be described as consisting of two up quarks with a positive charge of two-thirds each and one down quark with a negative charge of one third, altogether yielding a single positive charge. Similarly, one up and two down quarks would combine to form one neutron.

Despite their differences, all of Gell-Mann's quarks shared one common attribute: they were pointlike particles, with no apparent volume. This was an extremely important point for cosmology, because point particles, unlike the relatively massive protons and neutrons, could be densely packed into a small volume such as was thought to occur in the interior of a black hole or during the big bang itself.

In the less dense universe of today, quarks were theoretically confined to the atomic nucleus, except in high-energy accelerators where they could be broken loose for an instant or two. Nonetheless, in 1981 William Fairbank, a highly regarded Stanford experimentalist, announced that he had found a free quark—or something else like one with a fractional charge—using a sophisticated apparatus in a basement lab at Stanford.

Fairbank's announcement set off a brief flurry of fractional-charge hunting around the globe. The competition was intense among the raiders of the lost quark. None of the other experiments—in England, Italy and elsewhere—was able to confirm Fairbank's results, and some physicists joked that free quarks undoubtedly were only to be found along the San Andreas fault.

"I don't believe Bill Fairbank has found a free quark," Murray Gell-Mann said during a conversation I had with him in Aspen a year or two later. "If he has, though, it has to be one of mine."

The expressions *color* and *flavor* had sprung from the fertile mind of Sheldon Glashow, who had been a postdoc at Caltech during the days that Gell-Mann had been working up his new particle scheme. By now Glashow was teaching at Harvard. Even before Gell-Mann's three quarks began turning up in the Stanford Linear Accelerator in the late 1960s, Glashow had predicted the existence of a fourth quark, which, in compliance with the new lexicon, he had christened charm.*

In the early 1970s Samuel Ting, an American physicist who was born in mainland China and lived his first two decades there, had commandeered the large accelerator at Brookhaven. Ting was an unusually intrepid and rigorous experimenter and was known as a harsh taskmaster, strictly forbidding food and drink or even idle chitchat among his assistants during their long shifts at the controls of the accelerator. Despite Glashow's prediction, everybody in those days assumed there were only Gell-Mann's three quarks. Ting was merely attempting to confirm their existence and properties.

"We weren't looking for more," Ting recalled during a conversation with me in Hamburg, Germany, in 1983. "But we were taking a vast amount of data, and suddenly we began seeing a little blip on the computer projections. It could mean only one thing, that we had found a new particle."

In November 1974, Ting visited Palo Alto where he met with Burton Richter, an experimental physicist at the Stanford accelerator. Richter told Ting he had found the same new particle.

"I rushed back to Brookhaven to get my discovery into publication at the very earliest opportunity," Ting said. "I was able to beat Burt by one day."

No matter. Two years later the two ended up sharing the Nobel Prize for the discovery of Glashow's charm. The quark had appeared in both accelerators as a meson, an extremely short-lived particle consisting of the quark and its antimatter opposite, united in a momentary embrace before annihilation.

The November Revolution, as the Ting-Richter discovery came to be called, had not only shattered the simple and elegant concept of a three-quark universe, but it also, curiously, crushed the last resistance to the quark model among those theorists still holding out. In 1977 Leon Lederman and a team at Fermilab found a fifth quark called the bottom or, sometimes, beauty. In 1984 CERN announced that its colliding beams of energized protons and antiprotons had begun producing evidence of a sixth, the top, also known as truth.

By the early 1980s Gell-Mann's quark revolution had led to a new synthesis of what were believed to be the most fundamental units of nature. What Gell-Mann had brought about was little short of miraculous.

---

*The trend in fanciful particle names was not greeted with universal delight; at CERN I once saw a parody of the new physics lexicon written by John Ellis that referred to "winos and binos, squarks and sleptons."

A metaphor could be a football game played with an invisible ball in which all the players, the officials and the crowd seemed to run around or jump up and cheer without reason. To the uninitiated, the game would appear to be an exercise in futility. Then someone more intelligent would come along and hypothesize the existence of a ball and rules. At once the other observers would recognize the ordered pattern appearing out of what but a moment earlier had been utter chaos.

The overall pattern in particle physics that emerged following Gell-Mann's quark hypothesis was called the standard model. The model showed that the basic constituents of matter were twelve fundamental entities: the six quarks and six other particles in a class known as leptons, from the Greek word for light and swift. The most recognizable of these leptors was the electron; another was the muon, discovered in 1937 as the major constituent of the cosmic radiation continuously bombarding the Earth.

A third lepton was the tau, or tauon; discovered in 1977, it was like the electron, but about 3,500 times more massive. The additional leptons were three kinds of neutrinos, a name coined by Enrico Fermi since they carried no charge and were so small that they seemed to be massless.

Of the six kinds of particles in each of the two classes, only three—the up and the down quarks and the electron—played any role in normal matter here on Earth. The four remaining quarks, along with muons, taus, and the three neutrinos, existed only under highly unusual conditions such as those inside a high-energy accelerator or, supposedly, in the big bang. If these nine particles disappeared tomorrow, only physicists would notice the difference.

Science had made progress. Twelve kinds of matter (twenty-four, if you counted their antiparticles) were certainly more manageable than a hundred or more.* Beyond these, there was only gauge particles, a class of entities that carried the forces between the quarks and the leptons.

Variously called gauge particles, bosons or gauge bosons or, sometimes, gluons, these force-carrying entities existed for an instant within a fraction inside a fragment of a second. But during their vanishingly brief dance between life and near-instantaneous death, these force carriers performed the Herculean task of maintaining atomic structure, and thus the universe.

A distinct family of gauge particle, real or hypothetical, was responsible for the manifestation of each of the four forces—gravitation, the

---

*A few physicists, including Abdus Salam, the Nobel laureate from Pakistan, believed that surely there must be more fundamental particles from which both quarks and leptons are made. These hypothetical entities have been given names such as quinks, preons or rishons, and so forth. For the moment, quarks and leptons provide the paradigm for most physicists.

weak nuclear force, electromagnetism, and the strong nuclear force, in order of ascending strength—known to be at work throughout the universe; three of the gauge particles were known with certainty to exist, while the fourth, which carried gravitation, was entirely hypothetical. The forces gave structure to matter. Without them, there was just empty space.

The W and Z particles, which Rubbia began capturing in 1982, were gauge particles in a class called bosons responsible for transmitting the weak nuclear force that caused radioactive decay in the nuclei of atoms of certain heavy elements such as uranium. This was the force that was responsible for the nuclear reactions that powered the sun and the stars. Without the weak force, the stars might still exist, but only as cold, dark objects.

Before their detection, the W's and Z's were purely hypothetical; according to theory, they were unusually heavy, but existed for less than 10^{-18} second,* more often than not spending their lives entirely within the atomic nucleus. Rubbia's feat of detecting the "little beasts," as he was fond of calling them, was simply stunning, requiring all the technology and money his formidable talents could muster.

CERN built a special $20 million electronic detector specifically for the job. Big as a house, the detector was a vast network of coaxial cables and battleship-gray steel plates squatting over a section of the main accelerator where protons and antiprotons, magnetically whipped up to nearly the velocity of light, collided in a spray of debris. The charged particles liberated by these collisions streaked through the detector's gas atmosphere, generating trails of minuscule electric signals. These tracks appeared on computer screens as little V-shaped patterns, the arms showing where one had decayed into two or more others. But the W's and Z's themselves never appeared on the computer monitors.

"They didn't last long enough. But decay products from W particles, for instance, fly predominantly forward," said Rubbia during a conversation with me standing by the great detector one day. "When we began seeing this, we knew we had one."

The weak force imparted by Rubbia's particles was responsible for breaking down neutrons in the nuclei of a radioactive atom into a proton, an electron and an antineutrino. To borrow from a simile created by British physicist Sir Denys Wilkinson, imagine two skaters on a frozen pond. As they toss a ball back and forth between them, they recoil slightly from one another with each throw.

This was similar to what happened in radioactive nuclei. The skat-

---

*This is a billionth of a billionth of a second, a fragment of time a billion times smaller than the world's finest atomic clocks were capable of measuring.

ers were the constituents of the neutron, while the ball represented the W's and Z's. In the case of the nucleus of a uranium atom, the ball vanished with each throw.

The strong nuclear force, which binds protons and neutrons together, is 10^5 times stronger than the weak force, and binds protons and neutrons and their constituent quarks within the atomic nucleus. To use Wilkinson's analogy again, think of the skaters facing away from each other, tossing boomerangs over their heads to one another. With each toss, the recoil would carry the skater a bit closer to the other one. In the case of the strong force, the gauge particles that convey it are known collectively as gluons.

Although the strongest force in the universe, its range is only about 10^{-13} centimeters, about the diameter of an average nucleus. Paradoxically, the farther one quark is from another, the more powerful the embrace of the strong force, meaning that Bill Fairbank's isolated free quark really did have to be a true freak of nature.

About one hundred times weaker than the strong force, the electromagnetic force has an infinite range and affects all charged particles. The force's gauge particle, the photon, is responsible for light and other kinds of electromagnetic radiation such as infrared and ultraviolet radiation, X rays and radio waves. Electromagnetism is the force that draws negatively charged electrons to positively charged nuclei, thus allowing atoms to bind together into molecules and letting molecules themselves join together chemically.

Without electromagnetism, matter would not be solid; your finger would poke right through this page; but then, you wouldn't have a finger. May the force be with you indeed.

Gravity, the first force discovered but still the least understood, has virtually no effect inside the atom, being simply too weak in comparison with the three atomic-level forces. The electromagnetic field of an ordinary kitchen magnet, for instance, is stronger—over the space of a few

The Four Known Forces

Force	Relative Strength (compared with the strong force)	Range (centimeters)	Carrier
Strong nuclear force	1	10^{-13}	Gluon
Electromagnetic	10^{-2}	Infinite	Photon
Weak nuclear force	10^{-13}	10^{-15}	Boson
Gravity	10^{-38}	Infinite	Graviton (conjectured)

inches—than the Earth's entire gravitational field. Gravity, however, does have the rather big job of running the universe.

"Once you know about gauge particles, quarks, and leptons, you know everything there is to know," said Sheldon Glashow one summer day leaning casually against a door jamb at the Aspen Center for Physics. "That's all there is. There ain't no more."

He said this with a broad grin that hinted that even if he knew it he was not about to reveal the real secret of the universe.

Cosmic Nuptials

In the 1930s members of the two branches of the physics community concerned with the most fundamental affairs of the cosmos—atoms and galaxies—took to the floor to begin the long, slow dance that would, more than three decades later, lead to holy matrimony.

As is often the case, a simple question had initiated the courtship: Where did the energy for the stars come from?

At the turn of the century James Jeans, a British astronomical theorist, had constructed a basic model for the structure of a star: A mass of gas pulled together by its own gravity would have a dense, hot center. The energy radiating out from the seething core would make the star hot and bright, Jeans speculated. William Thompson, Baron Kelvin, one of the nineteenth century's most imposing intellects whose specialty was heat and electricity, took Jeans's idea and ran with it, estimating that a star such as our sun would burn out in about 10 million years.

Geologists already had found rocks on Earth they believed were far older than that, so they reasoned that the solar system's longevity had to be much greater than that. Because Lord Kelvin was so well respected, his calculations seemed to sound the death knell for Darwin's embattled new theory of evolution: 10 million years was simply far too short a time to allow for the biological changes required by evolution.

For decades astronomers correctly believed that a source of energy other than gravity must be firing the sun, but it was not until the late 1930s that Hans Bethe worked out the full details. A refugee from Nazi Germany, Bethe had landed at Cornell University in Ithaca, New York. He was so quick with numbers that he often left other physicists gasping in the wake of his problem-solving abilities.

Following a conference in Washington, D.C., organized by George Gamow, Bethe calculated in just a few weeks the mechanism by which the sun shone and the reason it had done so for more than a paltry 10 million years. His answer, nuclear fusion, was the first step down the aisle in astronomy's marriage with particle physics. Fusion was a process in which two light nuclei combine to make a more tightly bound heavier nucleus, releasing energy as they did so. (Fusion's opposite, fission, occurs during the breakup of a heavy nuclei to lighter ones.)

Bethe was able to show that four hydrogen nuclei could fuse into a helium-4 nucleus. This could account for the energy source of the sun as well as most of the other so-called main sequence stars. During the 1940s Bethe refined the mechanism of this fusion, which led to a Nobel Prize, while Gamow, Alpher, Herman and others began to work out the mechanism by which certain particles had been created during the first moments of the universe.

This was a highly speculative enterprise, because, during the 1940s, the theoretical instant of creation, which did not yet have a working title, was itself just a speck of imagination in the minds of a few physicists. A few more details of the stellar furnaces that lit the universe began emerging in the 1950s. Only a few years after coining the derisive nickname for the hypothetical creation of the universe, Fred Hoyle turned his unusual talents on a rare class of stars out of the main sequence, the red giants.

In 1954 he postulated that a special kind of nuclear reaction took place in red giants. Three helium nuclei would combine to form a carbon-12 nucleus but only at a specified energy level. Three years later Hoyle's friend and collaborator, William A. Fowler, in an experimental apparatus at Caltech, found to his great amazement the carbon-12 energy level exactly where Hoyle said it would be. For the first time, an astronomical prediction had been tested and confirmed in the laboratory.

The long courtship between subatomic physics and cosmology lasted until 1965, when Penzias and Wilson discovered the cosmic background radiation. This was a finding that everybody believed had finally joined cosmological prediction with electromagnetic technology. But the nuptials were not to be fully consummated for another fifteen years.

Almost immediately after the background radiation was found, particle physicists began taking a new interest in the problems of the early universe. In 1967 Andrei Sakharov, pioneer of the Soviet hydrogen bomb, analyzed the ratio between matter and antimatter in the current universe; he calculated that the present temperature of the cosmic background could be accounted for in a universe that had started out with equal numbers of particles and antiparticles.

During the 1970s theoretical physicists such as Hawking and Roger Penrose took the first steps toward what they hoped would be a single grand theorem that would explain the origin and evolution of the universe. In the meantime, particle theorists including Glashow and Weinberg were working to uncover a successful grand unified theory, a single explanation for the forces governing the interior of the atom.

Theorists had come to believe that their seemingly diverse efforts would be consummated in a grand convergence of cosmology and particle physics back at time zero, yet such a union still eluded them until 1981, when Alan Guth published his paper on the inflationary universe. Full and final union of cosmology and subatomic physics now seemed at hand. Particle physicists welcomed inflation because it offered them a quick cure for the seemingly terminal ailment of many of their theories: the prediction that the universe should be awash in magnetic monopoles left over from the first moments of creation.

You may remember Guth's theory required the existence of but a single monopole in the universe rather than billions upon billions of the unwanted particles. A single monopole swimming out there alone through the vast reaches of the cosmos was not very likely to place experimental constraint on anybody's theory. Who would ever be able to find that one, lonely little particle? The odds against it were simply staggering.

For their part, cosmologists were just as delighted with inflation as their particle brethren. Inflation had eliminated the flatness and horizon problems in a single theoretical stroke. Now cosmologists could concentrate on really important matters such as what had caused the big bang and how the universe had evolved afterwards.

For a few years the union of particle physics and cosmology seemed secure, even happy. By the mid-1980s Murray Gell-Mann observed that he now believed these two sciences had essentially become one. Individuals, of course, would still specialize in one or the other. MGM, for instance, would remain a particle theorist, while others would have a greater interest in astrophysics. But the unknown factors on both sides of the particle-cosmology equation had, for all practical purposes, become identical. The union meant, Gell-Mann believed, that humanity was on the edge of a great understanding about the universe.

"It is quite remarkable that a handful of beings on a small planet circling an insignificant star will have traced their origin back to the very beginning—a small speck of the universe comprehending the whole," said Gell-Mann as we talked on the lawn outside the physics center in Aspen one fine summer morning.

"It's a little like the ant contemplating the skyscraper, isn't it?" he said.

"No," he continued, "I think it's actually much worse than that." With a few quick scratches on a notepad, he calculated that the extrapolation scientists make in looking back in time in the direction of the big bang was more than one billion times as great as an ant looking up at the tallest building.

"But we are certain enough—arrogant enough, if you like—to think that our calculations are correct and that we are on the edge of a full understanding of the atom's nucleus as well as the origin of the universe."

The Ant and the Skyscraper

Little did Gell-Mann or anybody else realize in those heady days of the mid-1980s that, as each partner's weaknesses became glaringly apparent, the blissful union would soon be strained nearly to the breaking point.

PART III

No Guts, No Glory

*We hope to explain
the entire universe in a single,
simple formula that you can wear
on your T-shirt.*
—Leon Lederman

*The human race seems to love
long, impossible detours.*
—Charles McCarry

▼

CHAPTER **TWELVE**

ILLUSION
OF
SYMMETRY

There is in the world a great and yet
very ordinary secret. All of us are part of it,
everyone is aware of it, but very few ever think of
it. Most of us just accept it and never wonder
over it. This secret is time.
—MICHAEL ENDE

Despite
a few years of bliss and promise
for the future, the marriage of cosmology and particle
physics was not an easy one. Problems appeared on each side. But the
main hazard to harmony was that the partners played the game by different rules. Cosmologists saw gravity as the dominant force ruling the
universe. Particle physicists worked almost exclusively with the three
forces—the strong, the weak and the electromagnetic—that governed
the subatomic world. A number of theorists on both sides expected that
they eventually would be able to find an underlying relationship between
gravity and the other three forces.

"When the universe was very, very young and very, very hot, it is
highly possible that all the forces may have been one," said Sheldon
Glashow one summer day in 1984 as we talked in a seminar room at the
Aspen Physics Center.

"As the universe cooled, the forces split apart one by one—first
gravity, then the strong force, then the weak and the electromagnetic.
We hope and believe that we will be able to find the basic, hidden interaction which may have been present in the beginning."

Glashow had been a leading player in efforts to unify the forces ever
since his graduate school years at Harvard in the mid-1950s. In his thesis he had suggested that the electromagnetic force and the weak nuclear
force might actually be manifestations, under certain conditions, of the
same underlying force.

A few years later while working at Niels Bohr's institute in Copenhagen, he had refined the idea with a new mathematical scheme. Big,

Separation of the Known Forces　Many physicists believe that the four known forces separated and became individually distinct during the earliest moments of the universe.

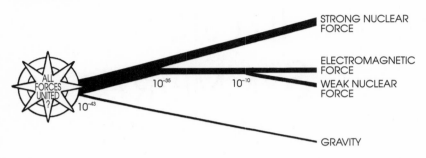

bright and outgoing, Glashow had the ideal personality for his important role in trying to unify the four known natural forces. As a middle-aged man he still had the charm and enthusiasm of a little boy looking through a telescope for the first time.

Nor was he the least intimidated when it came to putting forth his own new ideas or challenging the most established theories or even debunking new theories that he thought were more metaphysics than physics.

Searching for Symmetry

Behind the quest for unity that was at the heart of the new theoretical physics was the concept of symmetry. This soon led to another question: Why does mass exist?

In art and nature and geometry, almost everybody recognizes an object or a building or a design as symmetrical when it presents a similar appearance from different angles, or when it seems identical on both sides. Aesthetic appeal is a normal accompaniment—in a medieval cathedral, a Bach fugue or a snowflake.

Like so many of the West's most attractive ideas, symmetry was a Greek concept: from the word *symmetria* ("commensurateness"), a combination of *sym* meaning "together" and *metron* meaning "measurement." Such apparent simplicity contains numerous nuances relating to balance, proportion, consonance, harmony, beauty, agreement, order and, even, a sense of truth.

These qualities were inherent in the symmetry scientists began seeking in nature at the turn of the century. Physicists had begun recognizing that certain properties of physical systems were unchanged during apparent transformations. For instance, the effect of gravity on an object did not change the object itself. The size and mass of a cannon ball dropped from the Leaning Tower of Pisa remained the same during its fall as it had been at the top of the tower, even though its gravitational energy was transformed from potential to actual.

During the 1920s a mathematician named Emmy Noether at the University of Göttingen in Germany suggested that the conservation of momentum might indicate a symmetry or underlying simplicity in nature. Unable to lecture at the male-dominated institution, Noether nonetheless went on to make a striking assertion: There was no preferred location in which the laws of physics would work. They were the same everywhere.

Noether's concept of symmetry was to have a profound effect on particle physics; her idea was extended to mean that nature preferred simplicity. At the heart of the universe, beneath the apparent chaos of the cosmos and the wild, random movements of the submicroscopic quantum world were simple, perhaps even understandable, regularities no matter what changes seemed to occur.

The first natural symmetry in modern physics had already been discovered. Experiments by Michael Faraday in the mid-nineteenth century seemed to hint that electricity and magnetism, up until then believed to be separate phenomena, were somehow interlocked. In 1865 the genius of James Clerk Maxwell, the Scottish physicist who died at age forty-eight, produced a mathematical description showing that an underlying symmetry indeed existed; electricity and magnetism were the same thing.

Maxwell had concluded that visible light was merely one of many forms of electromagnetic energy, distinguished from the others only by wavelength. It was a splendid achievement. Yet nature was to prove more subtle than even Maxwell had imagined. Around the turn of the century Max Planck and others found that the full range of electromagnetic radiation could only be explained if its energy were thought of as consisting of tiny particles. The wave-versus-particle dilemma was not resolved until 1928 when Dirac invented the first known specimen of a quantum field theory, which was able to combine waves and particles into a single set of equations without even the suggestion of paradox (for those with the proper mathematical tools).

By the end of World War II nobody since Maxwell had seen a trace of further symmetry. A few years later Chen Ning Yang, the son of a Chinese mathematician who had grown up during the turmoil of the war and Mao's revolution, began to think that he saw a regularity in the strong nuclear force. Yang, then working at Brookhaven, believed such a regularity might exist because the strong nuclear force had almost exactly the same effect on entirely different kinds of subatomic entities, neutrons and protons.

Since the proton had a positive charge and the neutron had no charge at all, Yang thought that the strong force probably displayed a natural regularity, or underlying symmetry, in relation to charge. In other words, the strong force did not change even if the electrical sign did.

In 1954, the year Sheldon Glashow entered the graduate physics program at Harvard, Yang and Robert Mills, a Brookhaven colleague, codified this idea in the first modern gauge theory. They suggested for the first time that a basic symmetry, or regularity, might lie beneath all the forces.

Ironically, their paper, was plagued with its own irregularities in the form of inconsistent and uncertain mathematical formulations. But other physicists caught the drift, realizing as they absorbed the meaning of Yang-Mills's gauge theory that there could be a grand symmetry at the very heart of nature. Was that where the truth finally lay? If so, could it be found?

The notion was highly seductive, and thus dangerous. It could lead one down a primrose path. Like the idea that there simply had to be a fundamental piece of matter or that there was a final great truth about the universe, symmetry was very attractive. But its obvious appeal did not mean it was right.

About the time Yang and Mills began drumming up interest in symmetry as a fundamental concept, Glashow was looking for a thesis subject at Harvard. His adviser was Julian Schwinger, who had been one of the three authors in the late 1940s of an improved description of particle behavior known as quantum electrodynamics, or QED. It had occurred to Schwinger that under certain conditions the force of electromagnetism and the weak nuclear force might be manifestations of the same underlying interaction.

He suggested that Glashow work on the problem for his thesis. He did, finishing it in 1958. Although good enough to earn him his Ph.D., the paper did not resolve the original question. Two years later in Copenhagen, Glashow refined his work, coming up with a new mathematical approach that appeared to suggest that deep, natural symmetry would indeed emerge if the two forces were theoretically juxtaposed.

In a paper published in 1961 he proposed that another particle that he called the neutral Z boson, or Z^0, worked with the two W bosons—the W^+ and W^-, with opposite electrical charges—to convey the weak force between the particles it affected.[1] More startling, Glashow stated, the W and Z bosons would be indistinguishable from photons, which transmitted the electromagnetic force, in extremely hot environments. These environments might be produced in future accelerators.

Glashow's formulation suggested that the two forces complemented one another through an underlying symmetry. Unfortunately, his fledgling theory had a serious problem. It required that the W and Z force-carrying bosons be extremely massive, about 100 billion electron volts or roughly 100 times that of a proton.* If the two forces were, underneath it all, symmetrical, then the huge, though short-lived, W's and Z's should actually have a mass equal to 0, the same mass as the photons that transmitted electromagnetism.

In the language of physicists, the symmetry between the weak and the electromagnetic interactions somehow must have been broken by the mass of the intermediate-vector bosons, the W's and Z's. If there was indeed an underlying symmetrical force, though, these extremely heavy particles and the massless photons had to be the same entity, members of the brotherhood of the boson.

The flaw seemed fatal. How could such dissimilar particles be seen as one? Glashow's theory was viewed as somewhat interesting but unworkable. It all but drifted into obscurity during the next half decade.

Giving Mass to Mass

Glashow's fledgling idea was resurrected by Steven Weinberg and a Pakistani theorist named Abdus Salam. Weinberg's relationship to Glashow went back to the 1940s when they had been classmates at the Bronx High School of Science in New York City. They both had been members of the science fiction club and had begun discussing quantum mechanics there.

Both attended Cornell, with Weinberg going on to graduate studies at Princeton. While Glashow was extroverted and vivacious, and enjoyed fast autos, pool and cigars, Weinberg was introspective, bordering on solemn, a trait enhanced by a deep, somber voice. After Princeton Weinberg took a teaching job at Berkeley, then went on to London where he worked with Jeffrey Goldstone, a Cambridge theorist trying to uncover the electroweak interaction, as the hoped-for unity between electromagnetism and the weak force was by then being called.

Weinberg met Salam in London. An expert on the weak force who had achieved scientific fame in India owing to his remarkable mathemat-

---

*The concept of particles as tiny packets of energy follows from Einstein's formula equating mass and energy, $E = mc^2$. A proton, for instance, would have a mass of about 1.7×10^{-24} gram, or 938,300,000 electron volts. Particle physicists generally refer to particles in terms of these energy equivalents. One electron volt, eV, is the energy an electron would acquire as it was accelerated through a 1-volt electrical field.

Today's biggest accelerators can achieve a few hundred billion electron volts, or gigaelectron volts, expressed as GeV, or one or two trillion electron volts, TeV. This is just enough to begin producing the W and Z bosons, which were prevalent in the universe, according to contemporary theories, 10^{-12} second after the big bang when the same energy level was believed to have prevailed.

ical talents, Salam had begun graduate work at Cambridge at the age of twenty and had gone on to become an eminent theoretician. In 1964 Salam incorporated some of Glashow's ideas in a paper that took a stab at electroweak unification. Like Glashow's, his paper was incomplete and virtually ignored.

Everybody else was stuck, too. Perhaps they had been wrong all along. An underlying symmetry between the two forces simply might not exist after all.

Help was on the way from an unlikely direction. About the same time an unknown theorist named Peter Higgs was working at the University of Edinburgh on an idea that seemed only vaguely related. Extremely modest and reclusive, Higgs had a pudgy face that made him look like a cherub. Working on his own, he was investigating the possibility that all space was saturated by a field much the way an electromagnetic field saturated the region around an ordinary magnet. According to his theory, hypothetical particles would act as the agents of this undetected universal field.

About 1964 Higgs began noticing something extraordinary. If he added equations that were analogous to those of Salam's or Glashow's to his own field equations, certain particles would behave in an astonishing fashion. They would begin with zero mass at high temperature, then, Pacman-like, mathematically consume other particles in the field and emerge with mass as the energy level dropped.

If such were the case, then only the W and Z particles—and not the massless photon—would interact with the Higgs field, moving through it so slowly that it would be as if they were swimming upstream in molasses. At such a slow speed, far below that of light, they would acquire an effective mass; the slower the speed the greater the mass.

At higher temperatures, such as might be produced in a future accelerator or which might have occurred very early in the expansion of the universe, the interactions were such that the W's and Z's would no longer putter along but begin to pick up speed, and thus begin to lose their mass. They would now be one and the same as photons. This was an incredible statement. If it were true, it also meant that we are all living deep down in the depths of an ocean of Higgs particles on this low-energy planet.

As was his custom, Higgs in 1964 published his findings with little fanfare.[2] It looked at first as though nobody would ever pay any attention to the idea. Glashow went so far as to describe it as "terrifically loony."

In 1967 Higgs's work was suddenly recognized as immensely important. Independently, both Weinberg and Salam realized that if the Higgs field were incorporated into the equations they were working on, their problems would be solved: Electromagnetism and the weak nuclear force would be seen as one. Moreover, the Higgs field could also solve another major problem. It could give mass to point particles such as quarks and leptons. And this would answer one of the deepest questions in theoretical physics: Why does mass exist at all?

The invisible Higgs field was little short of miraculous, capable of nurturing massless particles, then giving them mass and sending them out into the cosmos on their own. The source of the miracle was the field's unique ability to preserve the universe's natural symmetry at extremely high temperature, such as might have occurred during the big bang, while being capable of breaking symmetry at low temperature. This was like water going from a liquid to a frozen state as the temperature dropped.

While a liquid, the water molecules consisting of two hydrogen atoms and one oxygen atom move rapidly, each pointing in an arbitrary direction. This is perfect natural symmetry, since no direction is preferred over any other. Lower the temperature below freezing and a subgroup of molecules aligns in one particular direction, forcing others to do the same. Symmetry is broken, since all the molecules in the crystalline ice now point in the same direction.

This was roughly analogous to how symmetry supposedly had been broken by the Higgs field. The phase transition that occurred as water went from a liquid to a solid was similar to what supposedly had occurred in the earliest instants of the universe. In the case of the universe, though, the lost symmetry had consisted of the elementary forces and their force-carrying particles. If Higgs were right, nature at its deepest levels might prove to be wonderfully symmetrical. But the world today was horribly unsymmetrical. What had happened in the meantime?

Apparently the Higgs field had done nature's dirty work as the universe had gone from high to low energy. This meant then, according to the calculations of Weinberg and Salam, that symmetry between electromagnetism and the weak nuclear force had been broken asunder the instant the universe had cooled to a temperature of about 100 billion electron volts. This corresponded, according to fans of the big bang, to an instant when the universe was but a trillionth of a second old. If such an energy could be reproduced in an accelerator (quite an impossibility in the mid-1960s), the beautiful and natural symmetry of the original electroweak force might be briefly recreated in the vanishingly brief lives of the W and Z particles.

The work of Weinberg and Salam had great aesthetic appeal to physicists and gained credence during the 1970s. In the meantime, Glashow, who had laid the foundation for the Weinberg-Salam model, found himself increasingly excluded from the credit. However, in 1979 the three were awarded the prize jointly.

In the meantime, the Higgs field was beginning to look like a major player in the history of the universe. Mysterious, unseen, and purely hypothetical, the field and its accompanying theoretical particles appeared to be responsible for no less than all the mass of the universe.

In theory, the Higgs particles acted like little magnets that would attract some kinds of metal more than others. Particles that had the greatest attraction would have the most mass. The Higgs field, though, had a much bigger and more immediate role: The electroweak interaction absolutely depended on the guess that it actually existed.

Finding the Higgs particle would confirm the basic concept of the underlying electroweak symmetry. Such a discovery might possibly allow the electroweak interaction to be extended to a higher-energy realm where the strong nuclear force and possibly even gravity could begin to figure in the equations.

The stakes were huge, the expense enormous. Thousands of people spent millions of dollars trying to track down the Higgs particle during the 1980s. Burton Richter was leading a team at Stanford, where he had rebuilt the linear accelerator with the specific intention of finding the Higgs boson.

Using a mammoth detector four stories high, Richter's old nemesis, Samuel Ting, headed the quest at the big new CERN accelerator. The hypothetical existence of the Higgs field was one of the chief arguments set forth for proposing to build in Texas the new American supercollider, a mammoth accelerator with a ring nearly the circumference of the 60-mile Washington, D.C., beltway.

There was, though, a slight problem. As of the early 1990s nobody had detected even the slightest trace of the Higgs field or its shadowy quantum particle. There was nothing more than a theoretical assurance that it would be found.

The situation, a few physicists realized, was a hauntingly familiar one, reminiscent of the nineteenth-century physicists' search for the ether. Not finding the Higgs field could be as just as revolutionary as finding it.

"Without the Higgs, Messrs. Glashow, Weinberg and Salam may eventually have to give back those little prizes they got in Sweden," said Paul Lecoq, a French physicist working with Ting at CERN.[3] He was only half joking.

GUTs of the Matter

During the same years that Glashow, Weinberg and Salam were bringing together electromagnetism and the weak force, cosmologists working with theorists from the particle side had begun to develop what they believed was a history of the big bang and the subsequent expansion of the universe until the present day.

Attractive to both the scientific community and the public because of its satisfying mythology, the big bang theory did seem to explain observations such as the apparent recession of the galaxies and the cosmic background radiation. Under the scenario worked out during the late 1960s and 1970s, the history of the universe begins at 10^{-43} second after the big bang.

This was the so-called Planck time before which physicists believed the normal laws of nature simply could not have governed the universe.* At that instant, physicists calculated, the temperature of the seething mass of space and time that was becoming the cosmos would have been an unimaginable (except mathematically, of course) 10^{32} degrees Kelvin—about 10^{19} billion electron volts, or gigaelectron volts.

Afterward, the universe passed through an immense range of temperatures to reach today's average temperature of about 3 degrees Kelvin.** The temperatures in between were a measure of the average kinetic energy of the particles, first in the hot, dense soup of the very early universe, then later as they coalesced into stars and galaxies during the subsequent expansion.

One apparent success of the big bang model was its ability to predict the number of elementary particles that seemed to exist throughout the universe from the number of protons—approximately 10^{80}—to how much of the mass of the universe was helium, about 25 percent. These numbers appeared to coincide with astronomical observations and were among the major pieces of data upon which the theory of scientific genesis rested along with, of course, the cosmic background temperature and the Hubble expansion.[4]

For most physicists electroweak unification meant that the universe's clock now had been turned back to just 10^{-12} (one trillionth) second following time = zero. This was the instant when, supposedly, electromagnetism and the weak nuclear force had split apart—at an energy level that could just be duplicated by today's accelerators. Moreover, the accelerator experiments of Carlo Rubbia, Samuel Ting and others had seemed to confirm that the theoretical speculations were correct.

Following electroweak unification, the next challenge already was clear: the creation of a grand unified theory, or GUT. This would bring the strong nuclear force into the fold of symmetry. With this, physicists would then have discovered the underlying symmetry of three of the known subatomic forces. Still out would be gravity, but everybody knew, of course, that gravity was a much more complicated affair. The principle in bringing the strong force into synthesis with the new electroweak force was the same: find an underlying symmetry, then determine how it

---

*The concept of time simply breaks down in units smaller than 10^{-43} when coupled with the spatial dimension of Planck's quanta.

**At room temperature, about 300K, the average energy of molecules is approximately 1/30 electron volt, eV.

had been broken in the short-lived era of cosmos-building right after the big bang.

Bold as ever, Sheldon Glashow already had begun attacking the GUT problem even before work on the electroweak interaction had been finalized. He had gone out of his way to become an expert on the strong force and had immediately recognized the awesomeness of the problem.

The theory of the strong force, called quantum chromodynamics or QCD, seemed to bear superficial similarities to the electroweak interaction, but it was far more complex. A single gauge particle, the photon, was the force carrier for electromagnetism, while at least eight gluons, the gauge particles for the strong force, were required for QCD. This raised the difficulty at least several orders of magnitude higher.

Undaunted, Glashow began working with Howard Georgi, a bright young theorist in the Harvard postdoctoral program. Meeting almost daily during 1973 they argued for hours over this new approach or that one. Nothing seemed to work. They tried one approach, then another.

One autumn evening after an unusually good afternoon session with Glashow, Georgi went home to relax. His mind wandered a little. Suddenly he was struck with the idea for a theoretical model that appeared to account for all the gauge particles in both the electroweak and strong interactions.

"I sat down, had a glass of scotch, and thought about it," said Georgi. He approached his new idea from another direction. It still worked. "So I got even more excited and had another scotch." [5]

The model he dreamed up called for a symmetry-breaking episode early in the history of the universe. In the case of Georgi's first GUT, this would have occurred at about $T = 10^{-36}$ second at an energy level 10^{15} (100 trillion) times higher than at the instant electroweak symmetry was broken.

In such an energized environment the gluons that carried the strong force would be interchangeable with the photons and W and Z particles of the electroweak interaction, and quarks and leptons would be the same, Georgi figured.

Symmetry would have been achieved all right, but the theoretical cost was great. There would have to be at least twelve new extremely massive particles to govern the transactions.

There was another even more horrible consequence, Georgi soon realized. It was inevitable that every proton in the universe would eventually decay. Its constituent quarks would simply degenerate. This would not occur for about 10^{31} years, to be sure. But Georgi found the idea strangely unsettling. It meant that as protons fell apart in the distant future, the stars and the galaxies would blink out one by one, leaving the universe utterly devoid of matter as we know it.

The reasoning behind such an awful prediction, however, was inescapable. If there were underlying symmetry between the strong and the electroweak force, then the strong force that kept protons intact and the weak force that caused certain kinds of radioactive decay ultimately had to be caused by the same interaction.

Like the radioactive atoms of uranium, every proton in the universe—all 10^{80} of them—was doomed. Glashow was less troubled than Georgi by such a prospect.

"We knew that the sun was going to burn out in a few billion years anyway," he recalled in a conversation with me years later in Aspen. "It didn't especially bother me that matter would start to fall apart a few billions years later."

Glashow and Georgi worked out a few details, then published in February 1974 the first grand unified theory, including its new heavy particles called X bosons and its prediction of the ultimate decay of matter. Called the SU(5) model, it drew some adherents and spawned a host of new GUTs.* Most GUTs incorporated the basic elements of the Glashow-Georgi model. Virtually all of them called for something like the X bosons as well as the dissolution of the proton and final decay of matter.

The first tentative steps on the road toward grand unification had been taken. The journey, physicists hoped, eventually would lead to an inclusive mathematical formulation that would not only include the three quantum forces, but also gravity—the wonderful and long-sought theory of everything.

Problems on the road to the great theory were apparent at once. It had not been a simple thing for Carlo Rubbia to hunt down the mediating particles for the electroweak interaction. It would be impossible to do the same for Georgi and Glashow's first GUT or succeeding ones.

Making use of the most complex technologies on Earth, the biggest accelerators at CERN and Fermilab could match temperatures in the early universe at what was believed to be about $T = 10^{-12}$ second after time zero. The superconducting collider planned for Texas would push this back another order of magnitude, to about $T = 10^{-13}$ second. The symmetry-breaking episode of the weak force and electromagnetism had occurred when the energy of the universe was a few hundred billion electron volts. The level theorized for the GUTS transition at 10^{-33} second after the big bang was about a million billion times higher still.

An accelerator capable of detecting the X bosons of the hypothetical GUT interaction would have to generate 10^{15} GeV. To be able to achieve such a high energy level using the most advanced technology imaginable would require a ring that would encompass the Earth and the nearest stars.

To test a theory of everything that would bring gravity into the fold

---

*In the model designation, the letters _SU_ refer to "special unity" while the 5 refers to the way fundamental particles are grouped in sets of five.

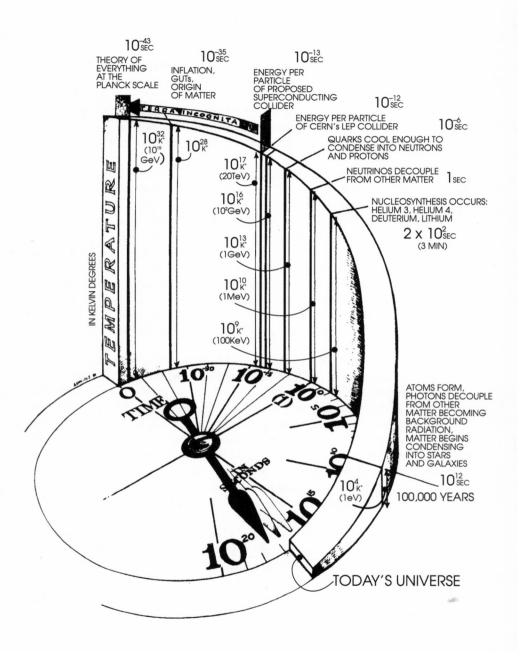

Thermal History of the Universe   Using even the biggest accelerators, physicists are unable to explore the terra incognita at the dawn of time.

a machine would have to reach an energy level of 10^{19} billion electron volts. This was a feat that would require an accelerator at least as large as the entire Milky Way.[6]

Such energy levels would never be duplicated on Earth. They had occurred but once, if at all.

GUTs proponents realized that there might be another way out. The various models predicted the eventual decay of protons. Perhaps such an event could be experimentally recorded. The proton lifetime called for by even the most optimistic GUT models was more than 10^{30} years.

This was much longer than the age of the universe, as it was then being hypothesized. Not many people would have the will or the means to watch a proton for such a long time to see if it would finally break apart. So physicists hit on another scheme: Why not watch 10^{30} protons for a while and see what happens?

In the late 1970s and early 1980s experiments sprung up all over the globe. The experimenters were enthusiastic and optimistic. It was, after all, a Nobel effort. If you could confirm grand unification, the prize would be yours along, of course, with the theory's inventors. Proton-decay detectors were constructed in deep mines and in tunnels to insulate them from cosmic rays and other interactions: in the Morton Salt Mine near Cleveland, in a zinc mine in Japan, a gold mine in India, a mine in South Dakota. A few were built in the alcoves of long auto tunnels in Europe.

One I saw in the early 1980s was two miles beneath the summit of Mont Blanc in the middle of the seven-mile tunnel between Italy and France. The air was rank with auto fumes and the odor of diesel. Pio Picchi, a lively Italian physicist who drove a white Porsche and preferred the cafe life of Rome to life in a cave, waited there for the death of a proton.

Picchi and his colleagues kept duty watches on the estimated hundred million trillion trillion protons in a 150-ton stack of iron slabs. Some 42,000 calorimeters, devices somewhat like Geiger counters, were fitted to the slabs, on 24-hour alert to pick up a minuscule burst of radiation that supposedly would be emitted during the death pangs of a proton.

If the GUT models were right, at least one of the protons in the mass of iron should give up its life for physics during the course of a year.* By the beginning of the 1990s, however, neither the Mont Blanc detector nor any of the others had recorded a single verified signal that was consistent with proton decay.

---

*Similarly, according to the GUTs, you can expect at least one proton in your body to expire during your lifetime.

A more promising possibility, some cosmologists believed, was to take a closer theoretical look at both the GUTs epoch of the universe and Alan Guth's inflation, which occurred hypothetically about the same time, at about 10^{-34} second after time zero. At Stephen Hawking's 1982 Nuffield meeting in Cambridge, theorists already had begun speculating that the appearance of the Higgs field at about the same instant would give space a net energy density that would continue until the force-carrying particles acquired mass and symmetry was broken.

During such a process, the universe's vacuum energy should have dropped to zero. This, it turned out, was exactly the false vacuum the cosmologists needed for inflation. Before inflation began at 10^{-35} second, the universe's energy density would be so enormous that such a vacuum would have no effect. At the slightly less energetic GUT level an instant later, the vacuum energy would become significant compared with the density of radiation.

This was indeed a wonderful moment in the history of the universe, at least in the minds of theoretical physicists. Inflation begins. Symmetry is broken. The flatness and horizon problems are solved. All but one of the magnetic monopoles disappear, as all the big problems of the early universe are magically washed away.

Of course, you could speculate all you wanted. Nobody would ever be able to test your theory.

Desperately Seeking SUSY

Despite the magic in their theories, none of the cosmologists could figure out how the infant cosmos at the time of GUTs and inflation had been able to expand into a universe containing the galactic clusters and superclusters being routinely discovered after the the mid-1980s.

The speculative Higgs field would have produced density fluctuations all right, but they would have been at least five orders of magnitude too dense. If such thinking persisted, the Nuffield theorists realized, today's universe would be filled with black holes in place of galaxies.

That problem, along with the failure of the proton decay experiments to turn up anything at all, indicated that the earliest and simplest models of grand unification were headed for an early exit. What was needed was a particle-cosmology model that would not produce black holes but that would result instead in the properties observed in today's universe, after inflation had been taken into account.

By the mid-1980s particle theorists began trying to skirt around these fatal difficulties by turning to a class of theories that had been proposed earlier with little notice. These fell generally under the rubric of supersymmetric grand unified theories, also known as SUSY-GUTs.

John Ellis and three of his colleagues at CERN in 1984 published a paper, "Inflation Cries Out for Supersymmetry." [7] They pointed out that the SUSY-GUTs could achieve this miracle of keeping the early density fluctuation amplitudes within the range required for proper galaxy construction while still performing the other tasks needed to convert the early universe into the one of today.

That was the good news. The bad was that the new version of grand unification led to an entirely new order of particles, putting partner particles of quarks and leptons, the stuff of ordinary matter, into a new class called fermions. Also included would be the partners of the charge-carrying gauge particles—the photons, W's and Z's and gluons.

By now the naming of new particles had become an existential enterprise. The partners for quarks and leptons were to be called squarks and sleptons. The so-called fermionic partners for the so-called bosons were christened with names like photino, wino, zino, gravitino, gluino and sneutrino, while the GUTs gauge particles would be called xino and yino. And the Higgs particle, of course, needed its SUSY partner: the Higgsino.

The new hypothetical partner particles provided enough latitude to allow the theorists to minimize the size of the quantum bumps in the early universe to whatever specifications they wished. Moreover, the breaking of the hypothetical supersymmetry through the Higgs mechanism would provide the universe with yet another phase transition with the possibility of still more magical transformations.

Theoretically, it would even be possible to detect a few of the lighter particles, such as the gluino, in accelerator collisions. This was because their mass was in the range of the W's and the Z's, which already had been found.

The technology, though, was unusually complex. By the early 1990s none of the new hypothetical particles had been discovered.

Accelerator at the Dawn of Time

Despite the success of electroweak unification, theorists still were unable to explain why the various particles had acquired the mass they had, or what really had occurred early in the universe or why the universe was its current size.

There were other equally vexing problems. Why did the universe have three spatial dimensions plus one of time? What caused the cosmos to expand? Why was there matter in the universe in the first place? How did that matter get formed?

What were the initial conditions of the big bang? Why did there appear to be no large concentrations of antimatter or a frontier of annihilation between matter and antimatter?

There was much speculation. But nobody really knew.

"We are far from such an understanding," observed Roger Penrose, a highly regarded mathematical theorist who worked with Stephen Hawking during the 1960s on the early universe.[8]

Acknowledging that there were serious failures on both sides of the particle-astrophysics equation as the new millennium approached, Penrose and a few other theorists were starting to see the light. On the astrophysics side, where the work of theorists like James Peebles and Jerry Ostriker was focused mainly on the later universe, observations had been running far ahead of the ability of theoretical models to account for them.

None of the theories was yet capable of fully accounting for the great walls, the giant attractors, the quasars that were too near the edge of the universe, or the huge voids.

On the particle side, some progress had been made. Yet there was an increasing sense of unreality among particle physicists. None of the models of the very early universe, the GUTs, the SUSY-GUTs or any other, had provided a description of events during the era when matter and the forces were separating in which any of them could place any real faith.

"We have no experimental handle on the very early universe," conceded Steven Weinberg, coauthor of the electroweak model and one of the first particle physicists to join the ranks of the cosmologists.

The big accelerators had confirmed the existence of Gell-Mann's quarks and the electroweak interaction. But verifying any of the GUTs models—as well as the various new theories of everything, whose speculations transcended even grand unification—was simply beyond the capacities of the biggest machines, and would remain so.

If it had happened at all, the big bang had been God's own accelerator.

CHAPTER **THIRTEEN**

CAT'S CRADLE

Time is the the best teacher.
Unfortunately, it kills all its pupils.
—attributed to HECTOR-LOUIS BERLIOZ

One
pleasant Thursday afternoon in
1987 I spent several hours with Roger Blandford, a
British theoretical physicist, in his office at Caltech trying to devise a
way to explain gravity for an article in *National Geographic* magazine.
We wanted to use everyday language without having to resort to the
equations of Einstein's general relativity.

"It's a very difficult mental image," Blandford finally admitted.
"From the perspective of general relativity, it is easy to say that gravity
is not a force at all. It is rather what happens when you put matter into
space-time." If this were true, it would seem to doom all the theoretical
efforts that were trying to link up gravity with the three atomic forces.

"General relativity and quantum mechanics really are two different
things," he said.

He tossed two books onto the desk. The first was on general relativity; the other on particle physics.

"One side sees geometry," he said. "And the other sees any force,
including gravity, as just something else to be quantized."

Gravity was the odd force out. Physicists felt sure about the electroweak unification, and they had made some progress on the grand unified theories (although without hope of validation anytime soon). But
gravity was something else again.

Not only did it not jibe with quantum mechanics at all, but intuitively it simply seemed different. Superficially, gravity was simple
enough; it was the force responsible for attracting conglomerates of matter to one another.

Yet, despite everything that science had learned since Newton,
gravity was still the most difficult of the forces to understand. Aside from
orchestrating all the universe, gravity also had subtle effects on Earth

163

never imagined by Newton. Geologists had found that it was weaker on a mountain top than in a valley, for instance, and stronger at the poles than along the equator; and there was a gravity trench in the Indian Ocean where it had less pull than any place else. Gravitational variations within the earth's mantle apparently govern plate tectonics and other geological movements. Oil prospectors now prospect for oil by looking for subtle local differences in gravity to find petroleum-bearing strata.

Nearly every mechanical device, from clocks to hydroelectric dams, borrows its energy from gravity. Most life forms are also dependent on it, space biologists have found. It governs the body's height, shape and mass, and it is essential for health throughout most of a person's life; then it makes a person old. Without it we become weaklings, victims of our own physiognomy. Soviet cosmonauts who spent close to a year in orbit suffered severe losses of bone minerals and muscle, as well as serious and detrimental changes in their cardiovascular systems.

These are gravity's effects, which are becoming well understood. Three centuries after Newton and three-quarters of a century after Einstein, though, the essence of gravity—the secret at the center—was still elusive.

A New Kind of Geometry

Thanks to general relativity, physicists had a broad geometric picture of gravity, but many details were still missing. A wild hunt by experimental physicists around the globe in the late 1980s showed just how sketchy the picture really was. It all began in late 1985 with the work of a theoretical physicist at Purdue University named Ephraim Fischbach.

His problem was snakewood.

Fischbach was sure he was on to something. If he were right, it would be one of the biggest discoveries in science since Isaac Newton saw an apple fall and then explained why. Fischbach believed that he might have come across evidence for a natural force counteracting gravity.

He knew it was preposterous. Such a force could only be science fiction or the idle fantasy of a wayward Midwestern theoretical physicist. Fischbach was filled with doubts. If true, his antigravity would change science, possibly forcing a reevaluation of Newton and Einstein. Even how we view the structure of matter and the evolution of the universe would change.

He had come across material he could not ignore. Through an analysis of some old data and a stroke of good luck, and with a few more recent findings in hand, Fischbach believed he had amassed a strong body of scientific evidence for his negative gravity.

To help confirm the discovery, he needed to know how gravity affected the substance known as snakewood. But neither he nor anybody else knew what it was; they only knew its name.

Fischbach knew he was treading dangerously. Gravity had a long and noble history dating back to Aristotle and was a cornerstone of modern science. Newton had first described it in his *Philosophiae Naturalis Principia Mathematica,* first published in Latin in 1687 and probably the most important single work ever published in the physical sciences. (Its significance was rivaled only by Darwin's *The Origin of Species* in the biological sciences.)

As the other forces had been discovered and, in the cases of electromagnetism and now the electroweak interaction, found to have an underlying symmetry, gravity persisted in remaining the outcast. Although the weakest of the forces by many orders of magnitude, it was the only force that humans could not control.

The other forces could be increased, decreased and occasionally even reversed. Not gravity. It could not be reflected, stopped or slowed. It always attracted. It never repelled.

At least until Ephraim Fischbach. The idea of antigravity had first crept—maddeningly—into his mind in 1979 when he and another physicist, Samuel Aronson, found experimental inconsistencies at the Brookhaven accelerator. In defiance of the normal rules of gravitation, particles called kaons (or K mesons) were drifting upward in the detector during collisions.

Fischbach and Aronson examined and reexamined the reams of data. They thought of every possible explanation, calculating all possible outcomes. Nothing worked.

On Halloween night 1985 Fischbach and Aronson got together to pour over the data one more time. The conclusion was simply astonishing. They didn't believe it themselves. The upward swing could only be caused by a previously undetected force of nature.

This was heresy. According to scientific dogma, there were four known forces and no more. A fifth force could be dangerous to one's career. Fischbach had a well-earned reputation for meticulous work and innovative ideas. Was he worried?

"You bet I was," said Fischbach, a personable native of Brooklyn who seemed only a little out of place in Indiana.

He and Aronson were so concerned they had made a mistake that they put the idea on hold for six years. In the meantime, they reexamined data on gravitation all the way back to Galileo. Ever since Aristotle, gravity had always occupied the minds of the best scientists. Galileo had been the first to do more than just speculate about its effects.

Rolling balls of different compositions and weights down inclined planes slowed them and made the effects directly observable; by these means he determined how fast objects fell. His experiments overturned the conventional wisdom of centuries. All objects, regardless of their composition, fell at the same rate of gravitational acceleration (now known to be 32 feet per second per second). If dropped simultaneously from a tower and if air resistance were ignored, a cannonball and a wooden ball would arrive on the ground at the same time.

Which was something that Galileo did not do. There was no evidence that Galileo had dropped cannonballs or anything else from the Leaning Tower of Pisa, Nicolo Beverini, a physicist at the University of Pisa where Galileo studied, told me as we stood on one of its precarious balconies.

"Galileo was one of those great, brilliant minds who was able to look at the world and see things a little differently from everybody else," said Beverini.

"He also kept meticulous records. He may have thought about dropping something from the tower, or imagined the effects. He probably did that. But he never wrote about it. And he certainly would have." [1]

Newton, one of history's two or three greatest thinkers, by any account, was born in 1642, the year Galileo died. Possessed of nearly terrifying powers of concentration, he invented calculus while still an undergraduate at Cambridge University. Some years later Baron Leibniz, the great German mathematician, came up with almost exactly the same system.

Newton had shown his invention to almost no one, his only confidant being Isaac Barrow, his old professor of mathematics. Barrow revealed Newton's work. Leibniz at first conceded they had been working on the same thing independently and simultaneously. But when the issue was pressed, Leibniz and his supporters maintained that Newton had borrowed the idea from Leibniz.

Johann Bernoulli, a Swiss mathematician, wanted to prove Leibniz's claim. He published two problems that required calculus for their solution, challenging anyone to solve them within a year. After a few months, Leibniz had solved one and was working on the other. Newton had not heard of the challenge. When he finally received the problems, he solved them both within twenty-four hours, and sent the answers to the Royal Society in London where the solutions were published anonymously.

"The lion is known by his claw," said Bernoulli when he saw the answers.

During an eighteen-month stretch before he was twenty-four, Newton, who apparently did start thinking about it after seeing an apple fall, worked out the laws of motion and universal gravitation, showing that the force that pulled the apple to Earth and the force that kept the moon in orbit were one.* In modern terms, he had discovered an underlying symmetry between two seemingly dissimilar physical interactions.

---

*History's second most famous fruit tree, or at least a clone grafted onto the stump of the original, which was felled by lightning, still stands at Woolsthorpe Manor, Newton's

Any two objects, no matter their mass, attracted each other at a rate directly proportional to their masses and inversely proportional to the square of their distance apart. Newton's new mathematical tool, calculus, explained why the apple fell straight down, instead of, say, sideways toward a nearby mountain or building. The gravitational mass of the earth was concentrated at a point lying at the center of the planet.

Newton sat on his finding for more than twenty years, until it was finally published after an accidental encounter. The current Astronomer Royal, Edmund Halley, was having difficulty calculating the orbit of a comet that appeared about every seventy-five years. He went to Newton for help and found that he already had done the calculations, but he had lost them. So he did them again on the spot. Halley recognized at once the value of Newton's accomplishment and offered to pay for publishing it.[2]

Gravity directed the comings and goings of all the planets and stars in the universe of Newton (he was unaware of the existence of galaxies). As for gravitation itself, it was a sort of cosmic cat's cradle of forces, with every star and planet tugging and pulling at every other celestial object across the great chasms of the universe beyond the Earth.

Newton's concept formed a basic guide to scientific thought until 1915 when Einstein overhauled it with a bold new vision of nature, general relativity.[3] For one thing, Einstein abandoned Newton's concept of absolute time and absolute space. The time measured by a clock now depended on the clock's velocity. This meant that there no longer was any way to determine whether events in different places actually occurred simultaneously.

And where Newton's universe had been clocklike, absolute, even stately, Einstein's pictured gravity as the very architecture of space and time. The universe was in actuality a single vast bed of gravity, not a hodgepodge of billions of attractive forces pulling and tugging at one another.

Such a concept grew out of the idea that in a perfectly uniform universe—one, in other words, that contained no matter—there would exist only time and a vast sheet of space, representing the possibility of gravity which would not yet exist. If you put matter—say, a star or a planet—into this universe, you would distort the sheet of space-time.

To use the most common analogy, this would be like putting a cannonball on a taut sheet of canvas. The dimpling effect would be gravity.

---

birthplace, about one hundred miles north of London. The tree is the only source of a small, foul-tasting, seventeenth-century apple called the flower of Kent.

Professor Rupert Hall, a leading Newton scholar now retired from London's Imperial College, told me that he was sure Newton really had seen an apple fall: "Voltaire first reported it, and Newton himself confirmed it at least twice."

If you added a smaller object such as a baseball, its dimpling effect would be less than that of the cannonball, and the smaller ball would roll toward the more massive one.

From such a perspective, it was perfectly reasonable to think of gravity as not a force at all. Gravity really *was* geometry. It was simply the normal behavior of matter in space-time. In terms of general relativity, gravity was a curvature caused by material objects in space-time. The universe itself, a machine fueled by gravity, was merely a consortium of the curvatures caused by all celestial objects pushing against the great canvas of space and time.

Unfortunately, the metaphor is only partially accurate, since it omits the fourth dimension—time—that is so crucial to understanding Einstein. Nonetheless, the message was a powerful one. Space-time told mass how to move, and mass told space-time how to bend.

There were hints, though, that space-time was the master and mass the slave in the symbiotic relationship that shaped the cosmos: The geometry of space-time not only could transport gravitational action from one piece of mass to another, but it also could carry gravitational energy through the void of intergalactic space at the speed of light.[4]

Like Newton, Einstein had used Galileo's work showing that all bodies fall at the same rate in a uniform gravitational field such as the Earth's. In the two decades prior to 1908, Baron Roland von Eötvös, a

Space-time A large object has more impact on the fabric of space-time than a smaller object, and thus more gravitational pull.

Hungarian physicist, carried out gravitational experiments intended to prove Galileo right.

Eötvös and his colleagues attached objects of different composition at opposite ends of a torsion bar that had been suspended at the center so it could swing freely. Then they measured how much the centrifugal effect of the Earth's rotation at Budapest's latitude would draw the suspended materials toward the north against the normal downward pull of the planet's gravity. The results were published in 1922 after Eötvös's death, and showed no apparent differences among a multitude of test materials.[5]

The Snakewood Chase

Three-quarters of a century later Ephraim Fischbach and Samuel Aronson took another look at the baron's findings. They were amazed by what they found. When closely examined, the Eötvös tests actually appeared to demonstrate that objects fall at slightly different rates according to their atomic composition. The more tightly packed the atomic nucleus, the slower the fall. Fischbach believed that Eötvös himself must have been aware of the discrepancies.

"That could be the reason he chose not to publish the results before his death," said Fischbach during a conversation with me at Purdue. "He was a very meticulous scientist."

Fischbach and Aronson had analyzed the test materials used in the original Eötvös experiments. Among them were asbestos, tallow, copper, water and platinum. They could not find one of the materials because they had no idea what it was: snakewood.

This turn-of-the-century name apparently had dropped from the vocabulary. Fischbach wanted to find it so he could have complete results. He had a hunch that it might be a dense, tropical wood. He sent inquiries everywhere, to chemists and lumber yards, to South America and to Hungary. No luck. One day in the library he read that nineteenth-century violin makers used snakewood for bows. He began contacting musicians.

On a visit to the University of Washington in 1986 Fischbach at last found his man: one Alexandrid Illitch Eppler, a balalaika player of Russian descent who played with the Seattle Symphony. By good fortune, Eppler kept a supply of aged snakewood in his basement for making instruments. He even had a piece he could trace to Budapest in the 1890s. Fischbach and his colleagues borrowed it, analyzed its composition chemically and plotted it on their density-gravity charts. It landed just where it should have. They had anticipated the fit, but not the incredible psychological boost it gave them.

About the same time Fischbach heard by chance about an underground gravity experiment near Mount Ida in the Australian Outback.

Frank Stacey, a geophysicist from the University of Queensland in Brisbane, had used a highly sensitive gravity meter to determine the strength of the Earth's gravitational field to depths of several hundred meters in two mine shafts.

The deeper Stacey went, the stronger the pull of gravity. He had expected this simply because he and his colleagues were getting closer to the center of the Earth's mass. They also had found something else, a force opposing gravity with about 1 percent of its strength and a range of a few hundred meters.[6]

Other measurements in mines and boreholes seemed to substantiate Stacey's results. When he learned of Fischbach's work, Stacey suddenly understood what he had been seeing all along. There was another force working against gravity, at least to certain depths.

Fischbach and Aronson went public in 1986 with their fifth-force theory. This heresy generated monumental controversy and set off a flurry of experimental and theoretical work around the globe.[7]

During 1988 and 1989, the years of peak interest, a count showed at least forty-eight experiments under way in the United States, Europe, the Soviet Union, Australia, Japan and India. Some of the biggest names in physics were involved—Rubbia at CERN; Val Fitch, a Nobel laureate at Princeton; and scores of others still hoping to make a name. Nobody believed the answer would come easily.

"Few experiments are simpler in principle, harder to put into practice, and so far-reaching in implication," observed Princeton's Dicke, the experimentalists' experimentalist who had contributed so much in so many areas of physics. The most direct way of testing Fischbach's theory was simply to repeat the Eötvös experiments with more modern equipment, Dicke suggested. Paul Boynton, a physicist at the University of Washington, a hotbed of fifth-force activity, decided to try it.

At first Boynton and his colleagues found variations they thought could be attributed to a fifth force.[8] A number of other experiments also had positive results. For a while it looked like the traditional laws of physics were being shaken at their roots.[9] Had Galileo and Newton, even Einstein, really gotten it wrong?[10]

Other seekers of the force, alas, found no evidence. Eric Adelberger, a colleague of Boynton's at the University of Washington, did a similar experiment. He saw absolutely nothing at all. At the Joint Institute for Laboratory Astrophysics in Boulder, Colorado, James Faller used a vacuum tube and a sensitive laser measuring device to compare the falling rate of masses of unlike composition à la Galileo, and came up empty.

In the absence of clear-cut experimental confirmation or denial, no theoretical consensus emerged. At one point Murray Gell-Mann, one of the arbiters of laws in particle physics, said he believed that it was fair to speculate on the possibility of a new force, with the proviso that all the calculations were correct. Other theorists did not like the idea at all.

In a conversation with me at CERN in 1986 Alvaro De Rújula, a dashing, blue-eyed Spaniard, predicted, "In a few years, this fifth-force rubbish will be gone, and we'll all be able to get back to good, basic physics."

When I spoke with him in Geneva two years later, he was a bit unsettled by some of the positive experimental results and had ventured into the fray with a couple of papers of his own.* If not a convert, he was now rather subdued in his criticism.

"In the absence of any two experiments with the same results, we can't really say anything scientific about it yet," he said.

John Wheeler, protégé of Bohr, friend of Einstein and the latter-day Mr. Gravity himself, was no fan of the fifth force.

"I think that it will prove to be a flash in the pan," he said during a conversation with me at an international gravity conference in Perth, Australia, in 1988. But then again, he said he really did believe in the merits of one of his favorite dictums: "The absence of evidence is not evidence of absence."

Two years later most physicists had come around to Wheeler's point of view. The evidence for Fischbach's new force simply wasn't all that substantial. But the frenzy over it, widely covered in the scientific and popular media, had demonstrated what a flimsy theoretical grasp physicists had on the still mysterious force of gravity.

Fischbach's fifth-force foray also inspired a rash of new work in basic gravity, as experimentalists set about to pin down better numbers for general relativity, non-Newtonian gravity (as the fifth force came to be called) if it actually existed, and Newtonian gravity itself. Oddly, the Newtonian constant of gravitation—known as g—was still one of the least accurately known of the universe's fundamental constants.

Life in a Warped Universe

One obvious difficulty in trying to understand gravity was that human beings were creations of gravity. Moreover, it seemed closely tied to human destiny.

"We are children of gravity," said Ralph Pelligra, a physician in charge of medical research when I met with him at NASA's Ames Research Facility in California in the late 1980s.

"You can look at a human life span in terms of gravity. A child has to struggle to learn to stand upright and walk against it. Then during our

---

*In one of these papers, "On Weaker Forces than Gravity" in *Physics Letters B,* 180:213, Nov. 13, 1986, De Rújula speculated on the force's range as "somewhere in between Mercury's distance to the sun and our distance to the moon," and conjectured: "The new force would make like bodies attract, and would therefore not constitute an elegant explanation of why protons have not been seen to decay."

prime of forty-five years or so we resist it with very little effort. Gradually, we reach the point where we start yielding to it again," he said.

"Sagging skin and organs, varicose veins, bent bones, arthritis, decreased vigor and immunity, failing hearts. They all come from the lost battle against gravity."

Gravity, which has controlled the evolutionary destiny of plants and animals from the brontosaurus to the baboon, hardly exists for insects. It actually presents little danger for any animal on the small side of a mouse.

"You can drop a mouse down a thousand-yard mine shaft and, on arriving at the bottom, it gets a slight shock and walks away," J. B. S. Haldane, the British biologist, wrote years ago in an essay called "On Being the Right Size." "A rat would probably be killed, though it can fall safely from the eleventh story of a building, a man is broken, a horse splashes." [11]

The earth's gravitational field is something you just take for granted, like the air (though nobody as yet has figured out how to pollute gravity). In fact, humans can be seen as gravity-dependent machines. It has dictated the size of our organs and our limbs. You can think of the spine as a cantilever and the arms as levers.

"Every bone, every muscle, every ligament, every joint, every piece of cartilage is precisely aligned to maximize mobility within the Earth's gravity field," said Pelligra.

If you don't think this is the case, Pelligra would suggest that you take a look at the shapes of the creatures that have evolved where gravity has little effect, in the water. What about a species of humanoids, having evolved on the moon with a gravitational field one-sixth that of Earth's? They might be twelve feet tall, but a little frail to try out for the National Basketball Association.

Jupiter has a gravity field 318 times that of Earth's. According to the Pelligra hypothesis, its inhabitants probably would look like pancakes with stubby little legs. What about the possibility of human-shaped beings anywhere else in the universe? In terms of gravitational evolution, you can forget it.

In the absence of some kind of new technology, they'll have to come at least as far as this solar system for us to find out. Space travel is incredibly tough on human beings, worse than NASA generally admits. While on assignment for *National Geographic* magazine in 1987 I visited the Soviet Union and had the good fortune of meeting several cosmonauts at the space biology facility in Moscow and at Star City, the training facility outside Moscow.

One former cosmonaut, Oleg Atkov, a physician who specializes in the cardiovascular system, was unusually forthright about the horrors of living in zero g. He had volunteered to spend eight months in orbit in order to test the effects of long-term weightlessness on the human body. He regretted the decision almost as soon as he was launched.

His first night in space, Atkov drifted uneasily through the interior of the Soviet space station, Salyut 7. It was impossible to sleep. His head, used to its own weight on a pillow, felt large and light. His face, bloated with blood no longer held down in his legs by gravity, felt puffy and sore.

When he closed his eyes, he became nauseated. His head seemed like it was spinning as if he had had too much vodka. He knew what was happening to him, but it made no difference. His inner ear, without gravity to direct it, was no longer able to sense up or down. He was miserable.

The Soviets planned to use data from the trip to assess the effects of weightlessness during a long trip such as one to Mars. Without intending it, Atkov also was functioning as a test subject for another of Einstein's great contributions to physics: the principle of equivalence. This stated that there was no difference between the effect of gravity and that of an acceleration. They were equivalent. A person falling will not feel his or her own weight; this simple idea (once you've heard it) led Einstein directly to general relativity.*

An orbiting space station falls continuously toward the Earth. Its rapid forward movement keeps it from falling. To Atkov the effect was the same as being inside an elevator in free fall down a shaft.

His body reacted profoundly. His muscles, no longer needed for supporting his body or lifting objects, began atrophying almost immediately and continued to weaken throughout the duration of the flight despite two hours of intense exercise daily.

"I could see them wither before my eyes," Atkov told me two years later at the space biomedical facility where he was then working. He was shortish, trim, intelligent and extremely pleasant-natured with the wide open face of a person in perpetual good humor.

"I became lethargic and extremely fatigued, far worse than I had expected." When he finally landed, Atkov's muscles had deteriorated to the point that he had to be carried away on a stretcher. "My one trip into space was more than enough," he said.**

---

*Einstein recalled his discovery of the equivalence principle: "I was sitting in a chair in a patent office in Bern when all of a sudden a thought occurred to me: 'If a person falls freely he will not feel his own weight.' I was startled. This simple thought . . . impelled me toward a theory of gravity." With the addition of mathematics, Einstein formulated general relativity.[12]

**I had the same sentiment about my one brief experience with zero g aboard a KC-135 aircraft, a military version of the Boeing 707 that NASA uses to train astronauts. The plane, with a near-empty fuselage padded like an asylum, flies in a parabolic curve that creates a sensation of weightlessness for about thirty seconds. This occurs when the plane reaches the top of a 45-degree climb, levels off and drops like a giant roller coaster. It was great fun the first dozen or so times.

During the trip, his underemployed bones had given up calcium through his kidneys and urinary tract (a situation that can have the nasty little side effect of generating kidney stones) and he had gotten thin and weak. He believed that bones might lose as much as a quarter of their calcium during a year-and-a-half trip to Mars.

Worse, though, was cardiovascular deconditioning. Outside earth's 1-g field, as much as two liters of blood and other fluids would rise from the legs into the thorax and head. This would trick the body into thinking it contains too much fluid.

"It has to take corrective action," said Atkov. "So it reduces blood volume. Blood constituents and hormone levels change. The heart gets smaller."

All in all, it was not a happy experience. It had taken him nearly two years to recover fully from effects of the trip.

"Never again," he said, would he even contemplate going into space. Atkov and more than a few other medical people think that a trip to Mars might turn out to be disastrous.

"What's the point in sending someone to Mars if he won't be able to stand up when he gets there?" said Dr. Harold Sandler, an aerospace physician, when I spoke with him in 1987 at the NASA's Ames research facility in California. Sandler believed that artificial gravity, which could be produced by a rotating spacecraft like the one in *2001: A Space Odyssey,* would be a requirement.

Space-time's Destiny

Human beings, as children of gravity, in the words of Dr. Pelligra, also embody time. According to general relativity, gravity was not a force like those operating on the atomic and nuclear scales. The reason was that gravity was a property of space itself as well as of time. Remember that in a perfectly uniform universe, one without matter, space would be entirely regular, homogeneous, unbent. Introduce matter, and space-time curves, warping to cause what we on Earth experience as gravity.

Time was intimately related, slowing down in the presence of an object of great mass. This was because light, one of the few constants in a universe described by general relativity and in a sense Einstein's master clock, would have to travel farther between two points and, in terms of general relativity, time would slow down.

In the *Principia* Newton had stated that time "flows equably with-

---

After thirty-three parabolas and a little over ten minutes of weightlessness, I along with most of the other dozen or so on board felt extremely nauseated and disoriented and understood why the plane was known as the Vomit Comet.

out relation to anything external." With special relativity in 1905, Newton's image of time as a river ran dry. Time was no longer an absolute flow. It was intimately bound to motion and space itself. Einstein declared that the measurement of time intervals was affected by the motion of the observer.

Two years later in 1907 Hermann Minkowski, a Russian mathematician educated in Germany and renowned for his contribution to the development of the theory of numbers, proposed a new kind of geometry that added time to the three ordinary dimensions of space, creating a four-coordinate system. It immediately caught on among physicists as a means of geometrizing special relativity and helped lead Einstein to general relativity.

Time, however, was relative. If you've ever been in a train in a station and noticed the train next to yours begin to move, you've experienced a related phenomenon. Which train is moving, yours or the other one? You can't tell until you see a third reference point. If this is another moving train, you'll become utterly confused. If it is the station platform, you can orient yourself.

In a similar way, time was relative for Einstein. But there was no final arbiter of time, no ultimate station platform.

During normal business on the surface of the Earth, the differences in time are little noticed because they are infinitesimally small. The relativity of time becomes significant only at great velocities. In an accelerator particles zip around the great magnetized rings at nearly the speed of light. In the cosmos galaxies, quasars and so on zip along through the cosmos at huge velocities.

The idea of relative time seemed simple enough, but the implications were staggering. A fast-moving clock would tick more slowly than clocks at rest. A clock aboard a spaceship traveling at 87 percent the speed of light would tick away the hours only half as fast as a clock on Earth, and a clock caught in the huge gravitational field of Jupiter would run more slowly than one on Earth.[13]

So time was relative, but that doesn't really explain what time is or isn't.

The late Richard Feynman demurred when I put the question to him in 1988. "Don't even ask me," he said. "It's just too hard to think about."

John Archibald Wheeler, who believed in exploring the most distant and difficult frontiers of physics, did like to think about time. He often pictured what would happen on the surface of a black hole where the gravity field is so enormous that not even light could escape. Time would simply stand still there.

When events are extreme—such as inside the atomic nucleus or at the beginning (or end) of the universe or at the edge of a black hole—our view of time becomes meaningless, Wheeler believed.

Time Near a Black Hole   Time may slow down or even stop altogether in the vicinity of an immense gravitational field such as that of a black hole.

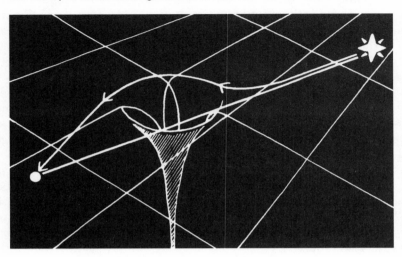

"This shows that time is merely a measuring tool, not an absolute flow or a substance," Wheeler said over lunch one May day in 1985 in the faculty restaurant at the University of Texas in Austin.

Without an event, time simply ceased to exist. It was Wheeler's view that time was only a secondary feature of nature, not a basic one at all. It was just another dimension, or something closely related to one.[14]

"Time is nature's way of keeping everything from happening all at once," was a graffito I saw on the wall of a cafe in Austin. If time were a dimension—something that gives meaning to events and the order in which they occur—this was literally true. The perception of time, then, would be a local phenomenon throughout the universe, quite literally a creation of the gravitational field of each celestial body and unique to any locale in the cosmos. Time on Earth was like time no place else.

This, of course, was how physicists looked at time. They saw it as a mathematical entity in their equations, only another dimension to be calculated. Whether time had a plus or a minus sign simply didn't matter.

Of course, everybody else, including off-duty physicists, looks at it a little differently. Every normal human being has an innate sense of the phenomenon of time that sometimes is called psychological time.* It is

---

*A number of psychological disorders are associated with a dysfunctional sense of time. A fascinating discussion of normal and abnormal time perception is in Göran Westergren's *Time: Experiences, Perspectives and Coping Strategies,* Stockholm: Almqvist & Wiksell International, 1990.

easily stated: As it goes along, you go along. How it goes along depends on the culture. In Europe and North America, time is linear. The past, present and future are arranged along a straight line without repetition of events or time itself.

"You cannot step twice into the river for fresh waters are ever pouring in upon you," wrote Heraclitus (535–475 B.C.). The Western concept of time apparently grew out of a Judeo-Christian tradition dating back to the Old Testament.

Each event on Earth or in heaven, occurring but once, assumed a special identity and a powerful significance: Genesis, the flood, the resurrection of Christ, the big bang. Such an idea did not allow for the hope of earthly reincarnation. But there could be life after death since it took place later in linear time (Pharaonic Egypt, with its surpassing concern for the afterlife, was more linearly time conscious than any other ancient civilization).

Western linear time was codified in Newton's theory of time as "true and mathematical," and later reinforced by the great discoveries of the nineteenth century: geological time was far deeper than anyone had ever before imagined, and biological time—evolution—had required millions, not thousands of years, a finding that made a joke of the famous pronouncement of Archbishop James Ussher of Ireland in 1648 that the universe had begun on Sunday, October 23, 4004 B.C.[15]

Time as a one-way track had powerful consequences for Western ideas of progress in science, technology, government, even religion. In the West today, the clock instructs us when (maybe even how) to behave. Each hour is a little bucket of time to fill. The clock says when to pour.

Edward T. Hall, a New Mexico anthropologist with a number of perceptions about time and culture, thinks such a scientific, logical, progressive sort of existence is controlled by what he calls "monochronic time." Life is time-ordered, scheduled, controlled, sometimes even fairly efficient. Such a fundamental concept is at the heart of the follies of our "you-only-go-around-once" culture: the cult of youth, the Type-A personality, a relentless consumerism always in search in something new.

In contrast, Hall observes, people in Asia, the Middle East, Africa and South America run their lives according to "polychronic time." Everything goes on at once: talking, eating, reading, playing, praying. Little attention is paid to the clock if there is one; schedules are flexible; and nobody minds.[16]

Almost everybody who lived before the scientific revolution of the seventeenth century (the first century, incidentally, in which people thought of themselves as inhabiting any particular century at all) very likely lived polychronic lives. Before then time was universally viewed as a circle, turning back or in on itself, all things possible at all times, a concept still central to Buddhist, Hindi and Taoist belief in which linear history is a fiction, all things returning again and again.

J. T. Fraser, a philosopher of time, thinks that an individual or culture's concept of time is its defining quality.

"A person's view of time is a way of discerning his personality," he once observed. "Tell me what you think of time and I shall know what to think of you." [17] We are defined by time, and time is what we define it as. Humans have internalized time so powerfully that it has taken on a meaning of its own. We run our lives—and our science—by numbers on the clocks and calendars of our own creation. How hard is it to look at time as pure abstraction? Most of us already have begun to anticipate Saturday, January 1, 2000. But in the grand scheme of things, it will be just another day.

Time and gravity are an intimate pair. Fischbach's work demonstrated that gravity has yet to be fully understood. The same is true for time. Although physicists might see it one way as they look out and survey the universe through their rose-colored Kuhnian spectacles, they are as much pedestrians in the great flow of time as everybody else when they take off those glasses and step into their everyday lives. And like anybody else, physicists are fully capable of mistaking one kind of time for another.

The confusion is understandable. Human destiny has been forged by gravity and sculpted by time. A creation of Earth's gravitational field, time is as much a projection of the mind as a feature of nature. We are not only within the field, we are a part of it. The apparent folly of fully comprehending the whole from within was, however, lost on those physicists who persisted in chasing the dream of the great theory of everything that supposedly was about to explain the entire universe to them.

CHAPTER FOURTEEN

THE
MASTER
OF
TIME

We dance 'round and suppose,
The secret sits in the middle and knows.
—ROBERT FROST

Several years before Stephen Hawking spoke at the winter 1986 physics meeting in Chicago, he had become intrigued by a remark from an L. P. Hartley novel, *The Go-Between:* "The past is a foreign country; they do things differently there."

Why? Hawking had wondered. Why was the past so different from the future? Why do we remember events in the past, but only dimly imagine what might happen in the future? And how was this all connected with cosmology? He began thinking there must be a relationship between the direction in which we perceive time to pass and the expansion of the universe.

"Scientists believe that the laws of physics govern everything that happens," he observed. "These laws do not distinguish one direction of time from the other any more than they prefer one direction in space over the opposite direction."

This could lead, he believed, to all sorts of science-fiction-like possibilities for people lucky enough to survive the traumatic transition from an expanding universe to one that was collapsing.

"Would they see broken cups gathering themselves together off the floor and jumping back on the table?" Hawking had wondered. "Would they be able to remember tomorrow's prices and make a fortune on the stock market?" [1]

Hawking decided that it should also be true that people would get younger and younger until they ended their days, not on the deathbed, but returning to the womb in the delivery room. In this bizarre world, every physical system would be affected by the direction of time. Rivers would flow uphill, and the water in them eventually would fall upwards to form clouds. Meteorites would fly off the ground, and their craters would close up as the meteors headed off into space.

These phenomena, naturally enough, would seem perfectly normal to the inhabitants of such a universe. Presumably, each one would lead a life counting down: first the years, then the months, then the days to his or her birth and disappearance into the uterus, back to the moment of conception and finally up the Fallopian tube to the ovary and then into a state of . . . what? Hawking's speculations hadn't gone that far.

This would occur because of what Hawking thought would happen to entropy in a contracting universe. Entropy was a measure of the degree of disorder in any system; in an automobile engine, a computer, a kitchen, a universe. The second law of thermodynamics, which was part of the theoretical bedrock of mainstream physics, dictated that, as time passed (in the direction we perceive today), the amount of disorder, or enntropy, increased because the energy required to create order dissipates.

Cars broke down, kitchens and bedrooms had a habit of becoming cluttered, the human body itself eventually degenerated, all seemingly of their own accord. In each case, an input of energy in the form of automotive maintenance, housecleaning or medical intervention could temporarily stall, or even reverse for a moment, the onslaught of entropy. Otherwise, disorder grew, with a portion of each expenditure of energy— to fix a car, clean a house or go to the doctor—radiating off into space, never again to be recaptured to do work on Earth.

"The second law of thermodynamics is based on the fact that there are many more disordered states than there are ordered ones," Hawking stated. He used the analogy of a jigsaw puzzle.

"There is one, and only one, arrangement in which the pieces make a complete picture. On the other hand, there are a very large number of arrangements in which the pieces are disordered and do not make a picture."[2]

What did this mean for cosmology? In the long run of billions of years, the expanding universe would dissipate all its energy and become utterly unable to fight entropy. Disorder would become total, the volume of space filled with nothing more than disembodied particles.

Hawking believed that if the universe eventually collapsed (a possibility, if there were enough mass to exert the requisite gravitational pull), the dissipated energy lost during the entropy process would begin to gather itself together, then gradually work to reverse entropy. Disordered states would gradually become more ordered.

The disordered state of a jigsaw puzzle really would gather itself into a complete picture, and a broken egg magically would seem to repair itself. Hawking's notion was based on the concept that the laws of physics are not generally constrained by the direction of time, unlike everyday life in which time seems to go only in one direction.

These physical laws would hold whether time flowed backward or forward. When Hawking first proposed the idea of a reversed-entropy universe, people in the physics community were astounded. Still, it came from Hawking, so it was worth listening to at least. Moreover, a few of the possibilities were quite intriguing.

"If time reverses when the universe collapses, it should also reverse inside a black hole," said Hawking. "So I have made life slightly easier for someone who jumps into a black hole." This meant that rather than being swallowed forever, he or she would be tossed out instead.

As the computer spewed out Hawking's latest thoughts about the direction of time that day in Chicago, the celebrated physicist himself sat almost immobile, slumped backward in his wheelchair on the podium. His head occasionally rolled to one side or the other as he and the others listened to the lecture he had written earlier and stored in the computer's memory, complete with his patented wry wit and inside jokes.

It was soon apparent to the physicists in the hall that Hawking was making a startling confession. He was admitting he had been wrong about the direction of time and was now actually retracting his own earlier ideas. "Work by Don Page [of Pennsylvania State University] and a student of mine, Raymond Laflamme, made me change my mind," Hawking's mechanical voice said. "I now think that the no-boundary condition implies that the disorder will continue to increase in the contracting phase. The thermodynamic arrow of time will not reverse." Hawking and a fellow theorist, Jim Hartle of the University of California at Santa Barbara, had worked out the no-boundary condition.

"The boundary condition of the universe is that it has no boundary," Hawking explained. "It would mean that the universe would be completely self-contained. It would not need any external agency to set it going or to choose how it began. The laws of physics would hold at every point. They would determine completely how the universe behaved."

According to the no-boundary idea, there would be no reversal of time during the contracting phase of the universe. Hawking said he had now come to believe in a model in which the universe was much more likely to contract to an extremely disordered state. The reason for his change of mind was that, like the jigsaw puzzle, the universe had the possibility of many more disordered states than ordered states. Voilà. The direction of time would not reverse; it would continue to point in the direction we've grown accustomed to, even when the universe starts collapsing on itself. So we could all relax, and breathe easier.

Hawking conceded during the speech that he had made an error.

His idea of people living their lives from death to birth in a backward-running universe could not possibly work. In fact, intelligent life could not exist at all in a contracting universe.

"Human beings have to consume food, which is an ordered form of energy, and convert it into heat, which is a disordered form of energy" he said. This meant that our accustomed direction of time was a prerequisite for life as we know it. Furthermore, his calculations showed him that, before the contraction could even begin, entropy would have proceeded on to completion.

Stars would burn out. Black holes could explode. Atoms would decompose, converting at least some of their mass into energy. Perhaps even the remaining subatomic particles would also decay (including the decay of protons as predicted by the various GUTs).

It was, as Jerry Ostriker had observed earlier, not a pretty picture. The universe would become a thin, homogeneous soup of light particles and energy as pure, total, and final entropy reigned. When Hawking's computer stopped talking, several physicists in the audience said they felt reassured to find that even Stephen Hawking could be wrong.

One confided to me, "If this time-reversal thing were really taken seriously, it would play havoc with our ideas about how nature actually operates." Actually, most physicists hadn't taken the idea very seriously in the first place, innately realizing perhaps that Hawking had carelessly juxtaposed time as a dimension, as it is used by physicists in their calculations, with time as ordinary mortals perceive it in the course of life on Earth.

The Dawn of Time

Physicists along with everybody else had listened, of course, because Hawking was Hawking. There seemed no question that he seemed uniquely qualified to make judgments about time's eccentricities. He had been the eldest of four children of an intellectual university family, their father a respected research biologist in tropical diseases. By the time he was eight or nine, he already knew he would become a scientist, although had decided not to follow his father into biology.

"It was too vague, too undefined," he said. Things could have been different, he recalled, had the more exacting field of molecular biology existed in the 1950s. Young Stephen had so many problems sticking to grammar school routine that his family worried he might not pass the entrance exam for Oxford.[3]

Stephen's father, an alumnus of University College, tried to pull strings to ensure his acceptance. But the father had underestimated the son. Stephen received a nearly perfect score on the physics section and performed so well during an oral exam that there was no question about admission, despite his unimpressive lower school record.

He was a popular, quick-witted student, and was coxswain for one of the university's eight-man rowing shells; he wore his hair long and couldn't get enough classical music and science fiction. His approach to studies was independent, undisciplined; his dons knew he had a first-rate mind, completely different from his contemporaries, and so they tolerated him.

He barely worked on physics. It was too easy. One day in a physics seminar, after reading a long solution he had worked out, he balled his paper up and tossed it across the room into a wastebasket. When it came time to graduate, Hawking needed first-class honors to receive a scholarship for the graduate physics program at Cambridge University, Oxford's ancient rival eighty miles away. During the vital oral exam an examiner asked him his plans.

"If I get a first, I shall go to Cambridge," Hawking replied. "If I receive a second, I will remain at Oxford. So I expect that you will give me a first." Which they did.

As a graduate student at Cambridge, Hawking began to show signs of becoming an inspired theoretical physicist. His friend and later collaborator, Roger Penrose, then a research associate in London, recalled that Hawking asked the most damnable questions, which defied conventional thinking and aimed directly at the weakest part of an argument.

Hawking was intrigued with theory. He had already taken his first look through a telescope one summer during a special course at the Royal Greenwich Observatory. One night Britain's Astronomer Royal, Sir Richard Woolley, asked Stephen to help measure the constituents of a double star. Hawking took a quick peek through the telescope's eyepiece and saw a pair of fuzzy spots going in and out of focus. He almost never looked through a telescope again.

At the beginning of his first year in graduate school, after returning from a trip to the Middle East, Hawking began losing dexterity and seemed to have a slight paralysis. It was hard for him to tie his shoes and, occasionally, difficult to talk. Nobody could diagnose it. One doctor thought he might have picked up a virus on his trip. Eventually the specialists decided he had amyotrophic lateral sclerosis, a rare and crippling disease sometimes called Lou Gehrig's disease after the great Yankee first baseman who died from it. The same disease claimed the life of David Niven in 1983.

Hawking's condition deteriorated quickly. Doctors gave him just a year or two to live. He began spending days and nights in his room reading science fiction, drinking by himself and listening to Wagner. Stephen's father asked his tutor, Dennis Sciama, one of theoretical physics' shining new stars, for special help on his son's dissertation.

Sciama declined. He knew Hawking was depressed. But Stephen would have to complete the work on his own. A few months later, as his condition seemed to stabilize, Stephen realized that death was not imminent. He began working on his dissertation again.

By the time he received his doctorate, he was in a wheelchair most of the time. Working as a research associate at Cambridge, he began collaborating with Penrose, then a young mathematician and theoretical physicist at Birkbeck College at the University of London, who already had established himself as one of the world's foremost mathematicians.

Penrose and Hawking set an ambitious course together: to find out if time had a beginning or could have an end. Penrose already had done some preliminary work on what they believed was a related phenomenon—collapsing stars. The history of any star, an average one such as the sun or a giant like Antares with a diameter the size of the Earth's orbit, was essentially a tug-of-war between the powerful outward energy of its internal nuclear reactions and the force of gravity that held the star together.

If a star's nuclear furnace started to burn down, gravity would begin overpowering the outward push of heat and radiation, and the star would collapse in on itself of its own accord. Penrose had wondered what was to prevent such a collapse from continuing forever? The star could crush itself down to an infinitesimal point containing all its matter, a single speck of infinite density and energy—a pure radiating little dot of hell.

No other theorist had been able to determine what would happen when the star reached such a point. This was a theoretical abyss, a place beyond the beyond, the end of the road, a place where space and time would simply disappear; it was known as a singularity. Here the equations relating to concepts of space and time, broken down in a theoretical thicket of infinities and useless zeroes, revealed nothing more to their masters.

In those years of the mid-1960s, most physicists believed that the singularities at the dawn of the universe, in fact, would turn out to be little more than mathematical abstractions with no real meaning, a metaphorical end of the road if not an actual one. A few years earlier, Penrose, in a remarkable mathematical tour de force, had shown that an endlessly collapsing star was more than just a theoretical toy. Such a star would actually end up as a physical, true and natural singularity.

Hawking had greatly admired Penrose's proof of singularity. Each recognized its ramifications. A star, of course, was one thing. The universe was something else. One of the problems theoreticians had faced was trying to figure out what would happen to particles as the universe contracted down to a mere point.

In 1963 a group of Russian theorists had proposed a theory calling for alternating contracting and expanding phases during the big bang that would keep particles from striking each other. Penrose and Hawking tried a simpler approach, looking at the way points in space-time would be causally related to each other.

This allowed them to avoid analyzing individual particles at the moment of the big bang as everyone else was doing. To do this they used the field equations of general relativity to look at the properties of space itself during the instant following the big bang. What they found to their great surprise was that the simplest solution seemed to be the most likely one.

Not only could there have been a singularity at the beginning of the universe, but an interpretation of the universe using general relativity seemed to actually _require_ a singularity.[4] The work revitalized theoretical work on what was then still a fledgling theory of the big bang.

Wizard of Black Holes

Soon afterward Hawking started thinking a lot about black holes. These were stars that had collapsed to such a degree that their gravitational fields had become so strong that not even light could escape. Among the strangest denizens of the depths of space, black holes had caught the public's fancy as few other celestial objects.

The evocative name, coined by John Archibald Wheeler, no doubt helped. Hawking believed black holes might enable him to understand the early universe where conditions were supposedly similar and also seemed to defy the known rules of nature. He thought that the death throes of a star might not be the only place a black hole could be born. His reasoning was that millions of minuscule black holes the size of protons could have been squeezed into existence by the monstrous power of the big bang. According to his calculations, the miniholes had been created within the first 10^{-20} second after the universe's supposed opening event.

The little black holes quickly caught on with the physics community. After all, such heavy little objects could even account for some of the mass needed to halt the universe's unhindered outward-bound journey that could end only in a cold, dark realm philosophically unacceptable to most physicists. Or, as some physicists suggested a few years later, for the dark matter in the galactic halos discovered by Vera Rubin.

In 1973 Hawking turned his attention to the behavior of matter that happened to be in the vicinity of black holes, big or small. As he twirled and juggled the geometrical hieroglyphics of black holes in his head, he formed a preposterous hypothesis. He didn't believe the idea himself and was privately embarrassed that it had come to him. The notion wouldn't go away. He tried to make it vanish. One evening he locked himself in a bathroom to work it out. There was no getting rid of it. The idea seemed stuck in his mind.

What he proposed was that black holes, in defiance of every known principle, emitted a steady stream of particles at their edge, a dividing line between night and day, between the known and the unknown, the real and the surreal, which he called the event horizon. He spent weeks more trying to make the damnable emission go away.

It wouldn't leave his mind. Finally he was convinced: Black holes, a creation of general relativity, could emit particles under certain conditions with the application of quantum mechanics. What allowed this to happen was Werner Heisenberg's uncertainty principle, the backbone of quantum theory.

A corollary of this principle was that space was never actually empty. It was, in fact, active and cluttered. Pairs of elementary particles such as electrons and their antimatter opposites, positrons, spontaneously flashed into being for a vanishingly small instant, then united to cancel each other out in a burst of mutual annihilation.

Hawking thought the energy for this instantaneous creation and extinction might be able to come from a nearby gravitational field. If such an event were to occur, say, at the event horizon of the black hole—the exact line across which even light feared to tread—one of the particles could be trapped by the black hole's enormous gravitational field. The other particle could escape if it were positioned just slightly farther away from the all-consuming event horizon. If somebody happened to see this, it would look like the surviving particle had actually been ejected by the black hole. Hawking was surprised and worried by the theoretical finding. Still, he went on. Possibly over a very long time, he soon recognized, the black hole itself would evaporate, having given up its energy piecemeal to the escaping particles. Hawking was usually cocksure of the creations of his intellect. But this time it was different. Evaporating and exploding black holes were so out of sync with the wisdom of the day that for weeks he sat on his findings.

He went over and over them in his head. Dennis Sciama urged him to go public. In February 1984 Hawking, still reluctant, was driven down from Cambridge to the Rutherford-Appleton Laboratory, a national facility in the unscenic flatlands west of London. The day he gave his paper he still wasn't sure of the results himself.

He titled it "Black Hole Explosions?" The question mark was a psychological escape hatch, an out in case he really was wrong. As the paper was read, most of the physicists listening had trouble following his ideas. Hawking became more worried.

After a few questions, the moderator, John Taylor, a professor of mathematics at the University of London, said, "Sorry, Stephen, but this is absolute rubbish."

Hawking still went ahead and published the paper in _Nature,_ the British science journal.[5] Later Dennis Sciama pronounced the paper "one of the most beautiful in the history of physics." For the first time, somebody had been able to demonstrate, at least mathematically, that black holes might not be cut off by an invisible barrier from the rest of the universe, but like everything else in the sky part of the continuum of space and time.

This was, still, theoretical speculation at best. By the 1990s observational astronomers had detected absolutely no trace of small black holes, emitting radiation or not.

The Quantum Wormhole

In 1988 Hawking published his wildly popular _A Brief History of Time: From the Big Bang to Black Holes._ He was then among the best-known scientists anywhere; by the time the book had run its course, he probably was the best known. The book had a two-part mission: a historical exegesis of cosmology and an exploration of his own thinking. In it Hawking acknowledged that he had made a terrible mistake about the direction of time.

"It seems to me much better and less confusing if you admit in print that you were wrong," he wrote. "A good example of this was Einstein, who called the cosmological constant, which he introduced when he was trying to make a static model of the universe, the biggest mistake of his life."[6]

Hawking's legend grew with articles about him, appearances on public television and radio, and a popular book about him; with the public nature of his physical and intellectual struggles, such as his declaration that time might be able to run backward, then changing his mind; with his well-publicized marital difficulties and eventual separation from Jane, his wife of more than two decades; with the suggestion once that physics could show us the mind of God while at the same time proposing a changeless universe that would have no place for a creator.

All the while, it was harder and harder to get a fix on Hawking as an individual. How good a scientist was he really, what kind of person? This was difficult even for his closest colleagues to know; like John F. Kennedy, veiled by his assassination and martyrdom, Hawking the person and the scientist was obscured from view by illness and his courage in the face of it.

"There is an aura around him, a spiritual atmosphere. He is going to end up as a saint," said one of his nurses, Indian-born Amarjit Chohan. There was no question that the atmosphere around him was colored with the light of another hue.

In 1987 a terrible event in Hawking's life had the indirect effect of enhancing the legend. While on a visit to CERN, he contracted a bad case of pneumonia. For a person in his already weak condition, nothing could have been worse. Near death, he was rushed back to England. Doctors there performed an emergency tracheotomy, a surgical procedure in which a tube attached to a respirator is inserted directly into the bronchial tube. Hawking survived, but at great cost.

He had talked before in a croaking whisper that only a few intimates could understand. Now he could no longer speak at all, something his family and closest friends had always feared.

For a few months Hawking seemed to lose his will to live. Finally he was outfitted with a computer-driven voice synthesizer. He could operate it with just two fingers. Small enough to travel with him on his wheelchair, the machine enabled everyone to understand Hawking now. Not only could he answer questions directly without resorting to a translator, but he could preprogram an entire lecture.

In another sense, though, the machine had removed Hawking another step away from everybody else. For one thing, it now took him five or ten minutes to compose a reply to a question. This could mean holding a roomful of physicists in solemn stillness while his computer slowly clicked out an answer.

People now seemed to speak very simply and clearly in his presence, as if he were deaf, feebleminded or speaking a foreign language. Most were probably trying to be precise, fearing that the amount of time it took him to answer would magnify the smallest error in their question. An aura of near-mythic proportions seemed to envelop Hawking. The eerie synthetic voice was utterly devoid of expression. His notoriously terse remarks took on an oracular quality that invariably precluded any kind of normal discussion.

"Students have to take what he says and go away and figure it out," said his old friend and colleague Roger Penrose, who had known Hawking from the earliest days of his illness. Penrose once had an excruciating three-day argument with Hawking concerning black holes.

"If he wanted to say something, he would make a noise on the computer. I would have to stop and listen, and I'd lose the train of my argument. It was extremely frustrating." The argument was never resolved.

In the late 1980s and early 1990s, with the question of time's direction laid to rest, Hawking pushed ahead in his quest for a complete understanding of what happened before and during the big bang. This led him to a vision of an infinity of interconnected and self-reproducing universes, an outgrowth of work on Alan Guth's inflationary scenario. Hawking began considering exotic and imaginary astronomical objects that he felt might lead him toward a a new approach for calculating fundamental quantities like mass and electric charge.

He hoped this work would eventually result in a quantum theory of gravity; this would combine in a single mathematical statement the dynamics of Einstein's general relativity and the horrible randomness of quantum mechanics. The physics community thought that such a theory would have to wait until a physicist specializing in the three subatomic forces and particles succeeded in devising a unified explanation of those particles and the forces, the elusive grand unified theory that Sheldon Glashow and other particle theorists sought.

Instead of waiting for revelation on the quantum side, Hawking pursued another strategy. In effect, he said, "Let's use what we do know about subatomic particles and their three forces to shed some light on the broader features of quantum cosmology."

Reviving an approach that was first explored twenty years ago, he applied the uncertainty principle—which he used with such powerful effect in the black hole emissions—to what can be thought of as the cosmic balloon. The universe could be seen as a kind of huge, rapidly inflating balloon, with the stars and galaxies moving around on the surface like points.

For this view we have to forget for the moment that we are part of the universe and imagine that we are somehow able to stand outside and look down at the surface of the vast balloon of space and time through a powerful microscope. At low magnification, the surface looks smooth. Zooming in, we see that the surface is vibrating like a foaming brook. Going in for a closer look, we see that the vibrations are, in fact, incredibly violent and chaotic, an unimaginable storm at sea—so wild and random that there is at least some probability that the surface at its tiniest scale could be taking almost any form at any time.

As far as Hawking was concerned, the essential thing was the probability that our furiously vibrating cosmic balloon would have developed somewhere, at some time, a bulge in its side similar to an aneurysm. This "aneurysm" could billow forth and, expanding indefinitely, create an entire new universe. Hawking called this new entity a baby universe.

"The field of baby universes is in its infancy," Hawking joked once at a seminar, then he typed into his computer, "but growing fast."

But, wait. If such a stupendous event had ever occurred, wouldn't we know about it? Hawking said no. The most likely size for the umbilical cord—what Hawking called the wormhole— connecting our universe to the baby universe would be the same as any typical quantum fluctuation at the surface, many times smaller than the smallest atom. If we could see it, this wormhole would look like a tiny black hole that flickered in and out of existence in an infinitesimal fraction of a second.

We would never notice it. Nor would we ever devise a way of testing such a hypothesis.

According to Hawking's speculations, a baby universe could easily behave like a full-fledged adult universe such as our own, inflating like a balloon of cosmic proportions at the other end of the wormhole. Eventually it could expand into a balloon billions of light years across. Nor would it have to be empty. The equations of elementary particle physics suggested that the sudden inflation of the baby universe could very easily produce an explosion of particles such as leptons and quarks that could one day populate the new universe with galaxies, stars, planets and, eventually, even life.

Hawking went further still. He believed that entire new universes could be coming into existence all around our own, that it was altogether possible that our own universe was formed this way. According to this hypothesis, our universe could be but a little quantum fluctuation on the great distended belly of another universe.

This was a bit disconcerting. Once the most important inhabitants of a world at the center of the known cosmos, now we human beings had been reduced to the status of the far-flung denizens of a minor, tangential blip on somebody else's universe.

A Labyrinth of Speculation

Hawking speculated that when and if the full quantum wave function of our universe became known, it would describe an infinite labyrinth of universes, splitting off and merging with one another—an endless sequence of cosmic creation and expiration. Such speculation had overtones of an Eastern epistemology that embraced recurring and eternal cycles of life and death.

Hawking had once called efforts to establish a relationship between modern quantum mechanics and Eastern mysticism "rubbish." His new conjectures were as surprising as his apparent embrace of the new cosmology. Initiated by Alan Guth's inflationary universe, the essential tenet of the new cosmology seemed to be what could be called a theory's auto-predictive power. Was the theory able to predict retroactively a universe that looked very much like the one we see around us?

This was a major flaw of the inflationary scenario, which purported to explain observed phenomena but was itself beyond testing by observation or experiment. In Hawking's own brand of the new cosmology, the most probable configuration for the universe was, indeed, a large, homogeneous cosmos expanding at a given rate and containing a certain number of stars and galaxies in specified proportions and distribution.

But what was surprising about that? After all, this was just the universe Hawking's theoretical framework, like Alan Guth's inflation, had been constructed to create.

Does theoretical speculation backed up with a set of equations say anything about the reality of nature? Hawking's new hypothesis was founded upon Einstein's general relativity, a good approximation of reality at the cosmic level, and quantum mechanics, which with its uncertainty principle denies a true view of reality on another scale.[7]

Could Hawking be proved right or wrong? What form might such proof take? Could such proof or denial be made even in the distant future? Hawking declined to say. But if it eventually were to prove to reflect reality, believed Hawking, then it would mean that the the laws of physics would apply everywhere: at the creation of the universe, at its end, through wormholes to universes beyond, everywhere, every time, in every branch of science.

"My proposal is the statement that the universe is a closed system," wrote Hawking in _A Brief History of Time._ "We don't need to suppose there's something outside the universe which is not subject to its law. It is the claim that the laws of science are sufficient to explain the universe."[8]

Hawking also touched on what it might be like for an astronaut accelerating toward a black hole while watching all eternity pass by outside his spacecraft in a single instant. Entering Hawking's presence seemed to give people the sense of a time dilation: an inexplicable slowing of natural processes that may have added decades to Hawking's life expectancy.

Watching the cosmos go by from a place a few degrees removed from the rest of us, Hawking himself seemed to be a metaphor for a lonely and curious species confined to a remote corner of the universe reaching out into the dark for understanding. Hawking had changed his mind, about the direction of time, about how time might end, about the boundaries of the universe—symbolizing the impasse that cosmology had reached by the 1990s. This was a period when the old paradigm no longer seemed to be working, and a new vision was not yet at hand.

In decades to come would historians of science look at the cold Chicago day in 1986 when Hawking publicly recanted his ideas on the direction of time's arrow as the defining moment? Was this the end of the beginning of the first age of humanity, in which we humans believed we would soon have the answer to the final mysteries of the cosmos, the all-encompassing equations that would explain what had happened at the beginning and how it would end, and why everything was the way it was in between? Was this the end of an era in which it was truly believed that we would soon have the great, elusive theory of everything?

When he was inaugurated as Lucasian Professor of Mathematics at Cambridge on a spring day in 1980, Hawking delivered an address he called "Is the End in Sight for Theoretical Physics?" He had been emphatic. The answer was yes. Theoretical physics would know the initial conditions of the universe by the end of the century.[9]

How innocent and self-conscious Hawking's title seemed a decade later. With cosmology in tatters and the big bang itself in jeopardy, the lesson during the intervening years had been a disillusioning one: that modern science very seldom attained its announced purposes.

Were the efforts of cosmologists to grasp everything about the whole universe as vain and feeble as they seemed? Some theorists already started wondering if, instead of awaiting revelation, they weren't detecting the willful tread of a wild and ancient god stomping, flattening, desecrating the hallowed temple where their finest work had been consecrated.

CHAPTER FIFTEEN

THE
METACOSMOS

So childish are we
that we wander around in time.
—BLAISE PASCAL

The great tragedy of science—
the slaying of a beautiful hypothesis by an ugly fact.
—THOMAS HENRY HUXLEY

A_n eerie haze of unreality enveloped the quest for the great theory that would explain everything there was to know about the universe by the early 1990s. None of the grand unified theories was on the verge of being put to experimental test anytime soon. This failing meant little to a cadre of theorists who had every intention of going ahead and concocting a theory of everything that would bring gravity into the fold, along with the still ununified three subatomic forces. Some of the ideas were so wild, though, that at times it looked like some theorists were playing mathematical games rather than making a bona fide attempt to comprehend nature.

It hadn't started out that way. In the mid-1960s Roger Penrose took one of the first cracks at devising a theory of everything almost by accident. One day while he was on a postdoctoral fellowship at the University of Texas, he was driving back to Austin after a weekend away. He was bored and let his mind wander. In a relaxed state, he suddenly had what he believed was a profound idea.

All at once a number of elements jelled in his mind, bringing together what he felt certain were the necessary ingredients for a new understanding of matter and the forces. With a little work, he believed, the idea would lead to a new theory of the universe. In that moment Penrose had conceived of the twistor. This was a mathematical entity that he believed was capable of defining all the subatomic particles and the four forces, including gravity, at nature's most fundamental level.

Penrose had the perfect mind to encompass something as revolutionary as the twistor. He had been to the scientific manner born, the middle of three sons of a British family whose intellectual roots went back for generations. His father, Lionel, was a well-known mathematical geneticist and inventor of geometric puzzles; his mother was a physician, and an uncle was a leading surrealist painter and confidant of Pablo Picasso. One of his brothers was a mathematics professor at the Open University, and another was a psychology lecturer and ten times British chess champion.

At Cambridge, where he received his Ph.D. in 1957, Penrose had worked under Dennis Sciama, the theoretical cosmologist who had inspired and guided so many others. The next year Penrose and his father coauthored an article called "Impossible Objects" in the *British Journal of Psychology,* in which they published a geometrical invention called a tribar and another design now known as the Penrose staircase.

They sent the article across the English Channel to graphic artist M. C. Escher in Holland, whose tessellated lithographs and their philosophical implications had made him a favorite of mathematicians. Escher loved the Penrose figures and immediately incorporated them in a pair of now-famous prints, *Ascending and Descending* and *Waterfall,* two of his most paradoxical visions.

Roger Penrose was one of the most mathematically and geometrically sophisticated individuals ever to work in cosmology. The range of his intellectual activities was simply staggering: algebraic geometry, differential topology, plane tilings, quasi-crystals and so on. Part physicist, part mathematician, conjurer of puzzles and puns, creator of optical illusions and impossible objects, Penrose was renowned for the creation of the famous Penrose Tiles. These were a pair of geometric shapes that could interlock with one another over and over so that they would cover a two-dimensional plane to infinity without ever repeating a pattern. Nobody had ever done anything like it before.

Short and handsome in a leprechaunish way, Penrose had a shock of wavy dark brown hair cascading over a face that seemed constantly bemused and a bit distracted. He was uncommonly genial, soft-spoken and well mannered for a leading light of theoretical physics. This was a world that was inhabited more by a kind of intellectual schoolyard tough—the brash, outspoken bully, the bright kid from a place like City College of New York or Harvard than by a genteel, British intellectual.

Penrose managed to get his ideas across because he seemed to have a certain feeling for what was right, what was beautiful, what piece to work into a puzzle. There was a musicality about his mathematics and his geometry, a special inventiveness, a unique intuition, a certain smell for the structure of the universe.

Working alone in 1965, Penrose had laid the groundwork for the twistor by demonstrating the conditions that would prevail as a large mass such as a star collapsed in on itself to the state of singularity. This is what would happen when the ordinary matter of a star—consisting of quarks and leptons—was crushed ceaselessly down and down by gravity, a cosmic vortex that had not yet been defined by modern physics. Using elegant mathematics, Penrose proved that, if it collapsed all the way down to a black hole, a star would have to encompass a singularity, a little blob of something and nothing all at once, a point where space-time, matter and forces simply ended.[1]

Four years later he and Hawking took up the same theme to show that there was nothing—mathematically, anyway—that would have prevented the universe from originating in a pointlike singularity.[2] This had been a stunning mathematical revelation for the early years of cosmology and helped establish both Hawking's and Penrose's careers. Afterwards Hawking went on to work on a quantum theory of gravity as a possible means of explaining the four forces and the underlying structure of the fundamental particles in a single set of equations.

Penrose's later approach to the same problem—the twistor—was, like the man himself, almost purely mathematical.

A Twist in Space-time

The idea that popped into Penrose's mind that day in Texas led him, over the ensuing years, to an unsettling, yet logically self-contained, vision of an eight-dimensional universe. Its fundamental entity was the twistor. He had extracted the name from the words *vector* and *spinor.*

The twistor was the building block of the Penrosian universe. It was an entity that he designed mathematically to encompass seemingly empty space as well as matter on the subatomic level. The twistor could account for all the subatomic particles, yet was smaller than any one of them. You could picture one in your mind by thinking of a pair of concentric doughnuts with a short shank of rope through its hole.

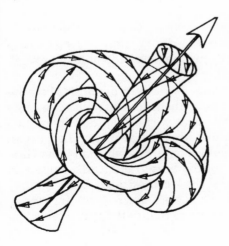

The Twistor   In the universe as conceived by Roger Penrose, the twistor is the fundamental unit of matter and energy.

The size and shape of an individual twistor, an entity that fell somewhere between an absolute point and a particle but was neither, determined its qualities. Certain combinations of twistor types could produce objects that seemed to behave like elementary particles. A single twistor of a certain type, for instance, could become one of the massless particles such as a photon or a neutrino.

This initially seemed like a marvelous accomplishment. It looked like Penrose had conceived of a building block of matter more fundamental than Gell-Mann's quarks. With an individual twistor or several in combination, Penrose might become capable of explaining not only quarks but all the other particles.

More incredible still, Penrose's twistor theory seemed capable of generating a kind of gravity theory at the subatomic level. This was something nobody else, including Hawking and his devoted band of followers at Cambridge, had come close to accomplishing.

Remember that in Einstein's theory of general relativity gravity was not really a force like the three that operated at the subatomic level. Gravity actually was a warping of space and time caused by the presence of large chunks of matter. In Penrose's twistor universe, gravity also was caused by mass giving rise to a curvature in the universe. But he had now shown how that warpage of space-time occurred deep down within the subatomic world itself, the shadowy domain that had never been penetrated by general relativity.[3]

Penrose apparently had brought it all together in a theory of everything almost before anybody else had even started thinking about it. Yet like the other theories that would soon follow, the twistor's early promise was not to be realized. Penrose worked on for years. The equations were complicated and difficult, even for one of the world's foremost mathematicians. When I first met him in 1981 at Oxford where he was Rouse Ball Professor of Mathematics, he seemed a little beaten down by the complexity of it all.

"It's proven to be much, much more difficult than I ever imagined," he said, standing at a blackboard covered with twistors. "Deep down, I feel certain that it must be right. But I just haven't been able to prove it. The mathematics are exceedingly difficult."

Two decades after the idea had first burst into his mind, Penrose estimated the calculations were still only about half complete; twistors only half explained the universe, which was worse than no explanation at all, as far as Penrose was concerned. Just as awful, he had begun losing his early adherents, a merry band of followers who had joined him at Oxford. The group included some of the best graduate students and postdocs from the United Kingdom and the United States. Early on they had seemed willing to follow the master puzzler down a winding, twisting road to an obscure but true mathematical reality they all believed in.

But nobody was willing to stick his neck out indefinitely for an idea that was not yet fully baked. Moreover, there were more than enough more conventional approaches for the newly disenchanted to follow.

As time went by Penrose was still unable to bring twistor theory, by the early 1990s more than two decades old, under control mathematically. He had grown weary and old trying. More and more he found his intellect looking around for new worlds to conquer. One of these new interests was the nature of computers and the possibility that they eventually would think like humans.

In his marvelous book, _The Emperor's New Mind,_ he brilliantly refuted this belief, held by backers of artificial intelligence. He believed and, of course, had demonstrated mathematically that quantum effects occurring in the human brain played a significant role in human thought and emotion. Such effects, he was sure, could never be duplicated in a machine. Because a machine lacked the quantum wiring of the mind, a computer could not have a temper tantrum or a brilliant flash of creative insight.[4]

No Strings Attached

With the supple geometry of his mind, Penrose had tried and failed to pull it all together. Other theorists would still try.

During the years Penrose was consumed with twistors, Hawking had trained the blackboard of his mind on the periphery of black holes, hoping to find the key to a quantum theory of gravity—and had found nothing. In the early 1980s he and a few others became excited by a fledgling theory they called supergravity.[5] A gravitational extension of the SUSY-GUTs, it was a multidimensional theory (the most popular versions had ten or eleven dimensions of space and time) that postulated particles as mathematical points. However, theorists working on supergravity were never able to make predictions that could be tested by observation or experiment. A bright, rising star for a year or two, supergravity dropped quickly below the horizon.

In the mid-1980s a new best contender for theory of everything appeared suddenly and brilliantly on the scene. The new theory's roots went back to 1970 when Yoichiro Nambu, a Japanese-American physicist, made a mathematical assumption that hadrons, the collective name for the particles that were governed by the strong nuclear force, could be interpreted not as point particles but as one-dimensional strings vibrating in twenty-six dimensions of space-time—twenty-five spatial dimensions and one of time.[6]

Nambu based his work on that of CERN theorist Gabrile Veneziano who two years earlier had postulated a theory writhing around in a similar multidimensionality. It had become apparent to other theorists that

Nambu's formulation applied only to the force-carrying bosons, however, and the idea fell out of favor. These were the years when electroweak unification was in ascendancy, and it appeared to most particle physicists that hadrons were made of quarks and not strings, anyway.

Four years later in 1974, John Schwarz of Caltech and Joel Scherk, a French theorist, revived Nambu's idea but modified it down to a mere ten dimensions and postulated strings as fundamental entities 10^{-33} centimeters in length. This was so small that, in relation to an atom, one of their strings was comparable to a single atom within the entire solar system.

The latest version languished for nearly a decade. The various GUTs still appeared to be a better approach, and strings were unusually esoteric even by particle physics standards. The mid-1980s were a period of theoretical uncertainty. Supergravity had failed, and the GUTs were beginning to look less promising. In the meantime, Schwarz had joined forces with Michael Green of Queen Mary College in London to come up with a new improved string theory that eliminated certain mathematical anomalies of the earlier versions.

Other theorists by now—in the early 1980s—were looking around for new bandwagons to board. Superstrings, the name Schwarz and Green gave their theoretical creation, began catching on. The name was a shortened version of "supersymmetric strings," unlike the cosmic strings of surpassing size or mythical power of the cosmologists.[7] Schwarz and Green had done their homework, creating a beautiful mathematical concept incorporating an underlying symmetry that seemed to be a way, theoretically at least, to unify all the forces. The new superstring theory was starting to look good.

"We've always had trouble fitting general relativity into any scheme," Schwarz said during a talk at the Perth gravity meeting in 1988. "However, with superstrings we find that not only does gravity fit, but it actually becomes inevitable."

Schwarz and Green apparently had created a totally new subnuclear universe in which their superstrings of vibrating energy, which lacked any internal structure, were the bottom-most entities. Always moving and oscillating, strings engaged one another in a free-for-all dance, swaying, bumping, sliding into one another in a process that Green and Schwartz showed mathematically could potentially create every type of subatomic particle—even the elusive and hypothetical little graviton.

Suddenly, geometry and quantum mechanics seemed to be as one at nature's most fundamental level.[8] But what about all those extra dimensions? They were a little pesky. Could anything be done about them?

Not to worry. In the 1980s Green, Schwarz and other physicists had rediscovered the theoretical work of Polish physicist Theodor Kaluza.

Back in the 1920s he had demonstrated that extra spatial dimensions could also be interpreted as special kinds of forces. Oskar Klein, a Swedish physicist, had then shown mathematically how these extra dimensions, or special forces, could be rolled up or compacted down to Planck dimensions of 10^{-33} centimeter; this would be so small that their presence would no longer be a bother. By the end of the 1980s theorists had fully reincarnated Kaluza-Klein—and extra dimensions were more popular than ever.

String Fever

Could anything possibly be wrong with superstrings? They seemed to do almost everything asked of them. Well, for one thing, the mathematics involved was all but impossible. In fact, like Penrose's twistor theory, exact solutions to the mathematical equations related to superstring theory were so elusive that by the early 1990s they still had not been found. Most theorists believed they were still years away.

Ed Witten was one of the best of the new breed of mathematical theorists who jumped on the superstring bandwagon. By 1991 he already had been working on the equations for several years. Like other string theorists he used a mathematical process called perturbation theory. This relied on approximations from one step to the next, with the equations becoming progressively more complicated.

It was somewhat like trying to find your way through a thousand-room mansion without a map. You would go into one room, then pick one of several doors. Then in the next room you would pick another door. Eventually you would hope to make your way to the master bedroom or the wine cellar or whatever other room you were looking for.

The difficulty of such a process mattered little to Witten. He was a believer. As far as he was concerned, superstrings were the thing. They represented the only possible supergrand unified theory. Witten had been a prodigy in math and science who, as an undergraduate history major at Brandeis University, hadn't even bothered to study physics. Even so, he applied to the highly competitive Physics Department at Princeton for graduate school. Despite glaring gaps in physics and mathematics on his transcript, he was accepted—presumably on the basis of sheer brilliance.

He had stayed on at Princeton after receiving his Ph.D., becoming a full professor at twenty-eight. After that, within a few years, he had won most of the major awards in physics and mathematics with the exception of the Nobel Prize. Like Margaret Geller, he had been honored with a coveted five-year "genius" stipend from the MacArthur Foundation (of which, incidentally, Murray Gell-Mann was a director). With superstring theory, Witten believed, physics had rounded a bend.

"We're now at the beginning of a process whose end we can't imagine," he said. He acknowledged that the end would not come for a while.

"No one invented string theory on purpose," he said. "It was invented in a lucky accident. By rights, twentieth-century physicists shouldn't have had the privilege of studying superstring theory." [9]

If it was a twenty-first-century theory that had fallen into the hands of twentieth-century physicists by mistake, what were they to make of it? The equations were difficult, and there were all those extra dimensions. Rolled up or not, they still had to be dealt with.

And why weren't our ordinary three dimensions of space and one dimension of time rolled up and compacted too? In other words, why had our ordinary four-dimensional universe been allowed to come into existence at all? Witten had been able to create a superstring theory that at least was able to eliminate the spatial nature of the extra dimensions. They still existed as undefined variables, however, and would have to be dealt with eventually.

The superstring zealotry of Witten, Green, Schwarz (who made a point of keeping an up-to-date list of new converts) and others who had quickly jumped on the bandwagon was notorious. For the converts, superstrings represented the long-sought theory of everything, the great secret of the universe, a grand statement explaining all the interactions of all the particles and all the forces including the intransigent gravity, a final theory from which all laws of physics, and presumably those of all the other fields of scientific endeavor too, would one day be derived.

"Superstrings may prove as successful as God," one of the theory's proselytes went so far as to declare, perhaps in jest. "He has after all lasted for millennia and is still invoked in some quarters as a Theory of Nature."

The fervor seemed to be based mainly on the theory's inner mathematical harmony, a place far from reality, where elegance, mathematical beauty and uniqueness seemed to be the main prerequisites for defining truth. According to the theory's critics, though, such fanatical insistence on mathematical purity had so far allowed superstrings to escape utterly the traditional confrontation between theory and experiment.

"The theory depends for its existence upon magical coincidences, miraculous cancellations and relations among seemingly unrelated (and possibly undiscovered) fields," observed Sheldon Glashow, superstring's most persistent critic. "Do mathematics and aesthetics supplant and transcend mere experiment?"

A few of the big names on the particle side did express an interest in superstrings, among them Weinberg and Gell-Mann. Others simply dismissed the concept out of hand. The late, great Feynman, a repository of insouciance whose utterances circulated widely through the theoretical community ("No one understands quantum mechanics," was one of his most famous) called superstrings "nonsense."

Clever as ever, Glashow denounced them wherever he could: at meetings, in talks, at symposia, and—most damaging of all—in a notorious article in *Physics Today* that he coauthored with Paul Ginsparg, a quantum specialist at Harvard who actually had worked on superstrings. They likened the theory to a medieval scholasticism in which pure thought alone defined nature. Superstring sentiments eerily recalled medieval arguments that design alone was proof of the existence of a supreme creator.

Eventually, superstring theory would be taught at "schools of divinity by the future equivalents of medieval theologians," they stated. "For the first time since the Dark Ages, we can see how our noble search may end, with faith replacing science once again." [10]

After more than half a decade of hard, driven work by some of the world's best mathematicians, superstring theory had yet to yield a single verifiable prediction or, in fact, to make any contact at all with experimental reality. Nor was that fact likely to change anytime soon, certainly not in the twentieth century nor perhaps even in the twenty-first. The mathematics were too difficult, too obscure and too dissimilar to those of conventional approaches, including the standard model of particle physics.

Witten, Schwarz and the others shrugged off such scathing criticism. They were, after all, seekers of the truth. Far more troubling for superstrings—the supergrand theory that would at last allow the final truth itself to be calculated—were its cosmological consequences. For the faithful, superstrings seemed more than capable of eventually providing its adherents with the opportunity to actually *calculate* what happened at the instant during the big bang when all the four forces were united.

This was, though, little more than wishful metaphysics. Superstring theory was operating far beyond the energy scales that could be tested in accelerators of today or even those of the middle to distant future. In fact, superstrings were out beyond even the GUTs levels in a never-never region of accelerator technology. To submit superstring theory to experimental testing would require an accelerator capable of attaining energies of the Planck level at 10^{19} GeV—a million trillion times greater than the Supercolliding Superconductor planned for construction in Texas during the 1990s.

Achieving such an energy level in an accelerator would amount to imparting the kinetic energy of a small commercial jet plane to a single proton. Even the most optimistic projections of the most adamant believers of continuing advancements in accelerator technology such as Leon Lederman, the vivacious former director of Fermilab, conceded a machine capable of such a feat could not possibly be built before the middle of the twenty-second century.*

The Metacosmological Frontier

The union of gravitational geometry with the three quantum forces supposedly had occurred at least once—during the big bang. Nobody, however, had been able to meld all the forces into a theory that could be tested. This did little, of course, to retard the speculations of theorists about what might have occurred during the big bang.

In a moment of grand speculation in 1969, a young physicist at Columbia University took the first stab at explaining why the big bang had occurred at all. Why, in other words, had the universe suddenly sprung out of nothing? During a seminar given by Dennis Sciama, the British theoretical star, Edward Tryon began daydreaming about the seething emptiness of quantum space.

"Suppose the universe is just a quantum fluctuation," he suddenly blurted out. Everybody in the room, including three Nobel laureates, roared. Tryon felt deeply humiliated. He was afraid to admit that he hadn't been joking. He tried to put the idea out of his mind. But it persisted. The critical element of the concept was simply too attractive: that the net energy content of the entire universe was exactly zero. This, of course, seemed ridiculous. After all, the energy content of the universe appeared to be enormous—starlight, all the energy bound up in all the matter of the cosmos, and the immense amount of energy that had been released during the big bang.

Sitting at home one evening a few years after his embarrassment at the Sciama seminar, he suddenly realized what could have happened. Suppose that all the apparent energy of stars and matter and everything else within the universe were considered to have a positive sum. Then the enormous quantity of gravitational force throughout the universe could be considered a negative.

It was just possible, Tryon thought, that the two sums could exactly counterbalance each other. In that case, each side of the energy ledger book would exactly cancel the other. The net energy content of the universe would indeed be exactly zero.

"If this be the case, then our Universe could have appeared from nowhere without violating any conservation laws," he wrote when he

---

*This projection is based on an empirical scale developed in the 1950s by Stanley Livingston of Lawrence Berkeley Laboratory who discovered a logarithmic relationship between time and technological improvements leading to higher accelerator energy levels: Available energy increased by approximately a factor of 10 every ten years. For nearly four decades Livingston's projection has proved to be remarkably accurate.

first published his new theory of genesis in the British science journal _Nature_ in 1973.

Tryon had been immediately struck by the concept because of what he called its "cold, precise, impersonal beauty." Within a few years the very physicists who had laughed at his original idea were studying it. What they found intriguing was the notion of the universe's colossal energy balancing act.

"We learned from Einstein that matter is a form of energy and our universe contains an enormous amount of energy," Tryon said. "But there is also another form of energy important to cosmology that acts, in some sense, in opposition to this mass energy, namely gravitational potential energy." [11]

This was similar to the potential kinetic energy possessed by a cannonball perched on the top of a tower. In the case of the universe, this was equivalent to the potential gravitational pull of every star and galaxy. This could be considered as a negative when it was opposed to the actual gravitational attraction of the stars and galaxies.

Tryon and most other theorists had faith in the idea that the universe was balanced just on the brink of being closed. In that case, the enormous positive mass energy of the universe just might be canceled out exactly by the negative potential gravitational energy. This would allow the universe to simply pop out of nothing without violating the laws of conservation of energy.

"In answer to the question of why it happened, I offer the modest proposal that our Universe is simply one of those things which happen from time to time," became one of Tryon's stock remarks. The idea was certainly provocative enough, and it won him the notoriety that would make him a regular in the popular science literature where he could usually be counted on to speak his mind in pithy pronouncements.

Others picked up the idea and ran with it. Vilenkin, the theoretically adventurous Soviet emigré, went even further in speculating on the initial moment of conception of the embryonic universe. As far as he was concerned, the minuscule bubble of space-time that would expand into the universe had tunneled out of potential space-time into actual space-time.

This tunneling process was a phenomenon that was believed to occur in quantum systems. The tunneling would take place under certain circumstances. Wave functions, which could also be viewed as particles, would break out of the quantum world and into the world of classical general relativity—a little quantum mole digging its way out through ordinary space-time.

Under such a scenario, our universe could have popped out of nothingness just like a tiny bubble of foam appears on the wave of an ocean. Many bubbles would appear, and most would disappear almost at once.

A few might survive, some even continuing to grow. Our universe was equivalent to one of these—a little bubble of space-time that happened to make it where almost all the others had failed.

There was no end to the speculative metaphysics, with other theorists playing out their ideas on the same theme. In 1978 four Belgian theorists proposed that the universe could have started out with the spontaneous appearance of two supermassive particle entities, one of matter and the other of antimatter, each weighing in the neighborhood of 10^{19} billion electron volts. These particles, in turn, would have stimulated the creation of more and more particles until all the matter in the new universe had been formed.

Three years later, in 1987, David Atkatz and the late Heinz Pagels, both theorists at Rockefeller University in New York, put forth a new theory suggesting the universe could have begun with a rapid change in the quantum-energy level of a multidimensional space that initially contained no matter. The sudden transition could have crystallized the tiny cluster of space-time into the ten dimensions required by superstring theory at the most fundamental level of matter and energy.

Nobody had yet explained why our universe, beginning as a little bubble on the great sea of space-time or as a multidimensional crystal, had grown so large. Neither Tryon's initial speculative outburst in 1969 nor the papers that grew out of it had addressed that problem directly.

The answer, of course, came in 1981, with Alan Guth's inflation, the ubiquitous cure-all for all the theoretical ills of the early universe.

The fledgling universe struggling for a toehold in the nothingness of the infinite void suddenly tore outwards in a hyperdrive of super-accelerated expansion, a 10^{-23}-second-long instant during which all radiation and particles were forged. Emerging from inflation with the matter and forces necessary for further construction, the universe expanded outward at a more dignified pace.

"Our universe was the ultimate free lunch," was the way Guth liked to put it.

Another ringleader in the new speculation was a Harvard theorist named Sidney Coleman, one of the wildest and wittiest theorists anywhere. He could be described as a cross between Albert Einstein and Groucho Marx. One legend demonstrated both qualities. Steven Weinberg was giving a seminar on quantum physics when Coleman burst into the packed lecture hall just in time to hear the Nobel laureate respond to a question, "I'm sorry, but I don't know the answer to that."

As he barreled down the aisle, Coleman burst out, "I do," then looked up and asked, "What was the question?" After hearing it, he answered it without a moment's pause.

Coleman had a problem with the cosmological constant. You may recall that Einstein had added this constant to the equations of general relativity in order to make sure that their solutions resulted in a static universe. This was a decade before Hubble. Nobody thought yet that the universe might be expanding, and Einstein feared that his theory might be seen as suggesting the universe was expanding or contracting (a possibility that was demonstrated mathematically a few years later by the Russian mathematician Friedmann).

Although only a number, the cosmological constant represented a tacit assumption that the universe contained a sort of antigravity force that would balance out the attraction of gravity on large scales. With the cosmological constant comfortably in place, the universe would be unable to go up or down in size like a balloon, at least according to general relativity.

After he learned of Hubble's discovery that the cosmos really was growing at an awesome rate, Einstein discarded the constant at once, declaring it the biggest mistake of his life. He now recognized, of course, that he should have been suspicious of a static universe all along. Had he kept the faith with his original theory, he himself would have predicted the Hubble expansion.[12]

Nobody thought much about the cosmological constant for five more decades—until inflation came along in 1980. Astronomers, after all, had looked out into the distant universe 10 billion light years or more and found absolutely no evidence for the repulsive force assumed by the cosmological constant. If the constant actually existed they would have seen some evidence in the movement of galaxies.

Such a repulsive force, if there was one at all, was so small that it could be virtually ignored. Yet many theorists realized that a serious problem persisted. If the quantum field theories such as quantum chromodynamics and quantum electrodynamics were correct, the emptiness of intergalactic space was not really empty. In fact, the void seethed with activity as vast numbers of virtual particles popped in and out of existence.

Since mass and energy were equivalent, gravitational effects from all this quantum activity should be observable—exactly as the cosmological constant had assumed. Yet no astronomers had detected a trace of such gravitational activity. The energy connected with the activity in the quantum fields spreading across the emptiness of space could be calculated fairly precisely, and was simply staggering: something on the order of 10^{120} times greater any observed effects.

Anybody can see that such a calculation is wildly wrong. Look outside. How far can you see? If the cosmological constant were as big as the equations indicated, the universe would be at most a mile or two across; space itself would be so curved that you literally couldn't see straight. Most likely, though, the universe never would have made it much beyond quantum dimensions. Almost immediately after the big bang the entire cosmos would have curled up on itself like a withered leaf.

This inconvenient detail had also led to a serious difficulty for the inflationary cosmologies of Guth and his followers. A spontaneous symmetry-breaking phase transition as postulated in the inflationary theories should have caused the vacuum energy of intergalactic space—the cosmological constant—to decrease by an amount that was far greater than the current cosmic energy density.

Each symmetry-breaking event was somewhat analogous to a transaction at a bank in which a contractor needed a tremendous amount of cash to continue construction on a project: If there were not enough money in his account, he would dip into his credit line. In the case of the inflationary universe, though, the credit limit—in the form of the cosmic energy density—had long since been surpassed.

How did the theory's proponents reconcile such an apparent bankruptcy? They had simply made the assumption that there had been a bigger credit line all along. Such an ad hoc adjustment in the initial conditions of their universe would ensure that it had an adequate cosmic energy-mass density in the first place to cover all future withdrawals.

Coleman disliked these assumptions. He also disapproved of the fact that the odds against an exact cancellation of the cosmological constant by gravitation were simply staggering. It was as if you spent millions of dollars over the course of twenty years without checking your income or your bank account. Finally you compared what you have spent with the amount that has been placed in your account, finding to your great relief that they balanced to the cent.

Something was wrong. Others had tried unsuccessfully to rectify this discrepancy. Coleman began looking around for the thread of an idea that would lead to a solution. In 1988 Stephen Hawking had proposed the quantum wormhole. This would be shaped like a tunnel and, theoretically, could be of any size, but, according to Hawking, one larger than about 10^{-33} centimeter would be extremely unlikely. This was some 10^{20} times smaller than a proton, in fact just about the size of a superstring. Hawking calculated that wormholes would be exceedingly short-lived, popping in and out of existence during the Planck time of 10^{-43} second.

In Hawking's highly conjectural proposal (even he acknowledged that it had not reached theory status), such a wormhole arising and disappearing nearly instantaneously on the space-time surface of our universe could protrude out into the void of voids beyond. This was the cosmic cul-de-sac Hawking called a baby universe. It could vanish as soon as it appeared, or it could expand into another full-fledged universe, perhaps even one like ours, perhaps an infinite labyrinth of interconnected universes.

Coleman seized upon the idea. As he studied the equations of

A Quantum Wormhole   Incredibly small and leading a very short life, a quantum wormhole briefly allows particles to pass through.

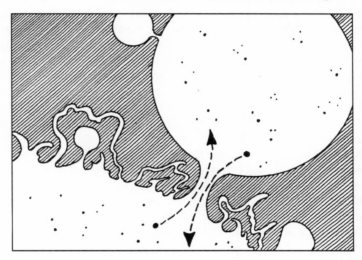

Hawking and the other wormhole theorists, he found buried deep within the mathematics a possible solution to the persistent and vexing problem of the cosmological constant. Coleman's idea, though mathematical in nature, was essentially this: The other universes connected via the wormholes to ours undoubtedly contained matter and energy that conceivably could interact with particles and energy in our universe through the wormhole during its brief life.

This was the necessary mechanism. Matter would have to be squeezed down somehow in order to pass through the wormhole, but this was not a mathematical impossibility. For Coleman the essential thing was that such a process would affect the way the fundamental constants in our universe were measured: The wormholes would fix the cosmological constant so that its value was zero, by exactly canceling all the virtual particles fluctuating in and out of existence in our universe's quantum fields.

In other words, an enormous feedback-adjustment-feedback loop would be at work, keeping the vacuum energy level constant and this universe on an even keel. The wormholes were nothing less than little cosmic tellers keeping the accounts straight.

When Coleman published his theory in a paper entitled "Why There is Nothing Rather than Something," there was a burst of enthusiasm within the theoretical community.[13] Not only did it seem to explain why the cosmological constant was zero, but it also could explain how general relativity, physics on the cosmic scale, might be connected to physics on the quantum scale.

One of Coleman's colleagues at Harvard, Steven Giddings, said he liked the feel of the parallel-universe-generating-wormhole-cosmologi-cal-constant-mediating theory. Others were doubtful. The theory made a number of assumptions that were not well understood, and the mathe-matics did not seem to prohibit large wormholes from popping up all over the universe.

Coleman, ever cheerful, said he felt he could answer most objec-tions. And even if he couldn't, he said, he was sure his wormhole idea was "probably not wrong." [14]

Right or wrong, the wormhole concept spawned a whole community of suppositions that were highly speculative at best—what could be called speculation squared. These ideas were far out on the fringes of science. Kip Thorne, a theorist who prided himself on being at the frontier of thought, and another colleague at Caltech, Michael S. Mor-ris, extended the wormhole idea to what seemed to be a logical conclu-sion.

Not only might the little cosmic wonder tunnels connect our uni-verse to others, but these little quantum passages might also lead to a distant region of our own universe. Better still, they could provide a tun-nel to another epoch of time.

Were they serious? In a paper they published in *American Journal of Physics* in 1988, they suggested that an advanced technology might even find a way of fishing a wormhole out of the quantum sea, and then expanding its size to dimensions great enough for an exceedingly brave human being to pass through.[15] This would be the case, they speculated, if the passage through the wormhole could be kept open by some kind of an exotic field or material capable of withstanding the ungodly pressures of trillions upon trillions of pounds per square inch.*

"We do not know enough today to either affirm or refute these diffi-culties, and we correspondingly cannot rule out traversible spacetime wormholes," Thorne and Morris, along with another Caltech theorist, Ulvi Yurtsever, wrote in another article.[16]

If you have hopes of getting in touch with the Eloi or Morlocks of H. G. Wells's *The Time Machine* (1895) anytime soon, though, forget it. The big problem with all such ideas, of course, was that there was no prospect for experimentally or observationally verifying them. Even more signifi-

---

*Four other theorists went so far as to propose a quantum time machine in an article called "Superpositions of Time Evolutions of a Quantum System and a Quantum Time-Translation Machine" which appeared in the June 18, 1990, issue of *Physical Review Let-ters* (pp. 2965–2968). Their idea was based on the habit physicists have of treating time and space the same way.

cant, the suppositions could not be disproven. This was a key tenet of the so-called scientific process that had been identified by the Austrian science philosopher Karl Popper in the 1930s. For a theory to be considered scientific, it had to be capable of being disconfirmed. Anybody could make up a theory. But until a way was shown to disprove it, it could not be considered a scientific theory, even if the theory had been proposed by a member of the scientific community with a dazzling reputation.

During the 1980s a new category of theory seemingly had appeared with the heading, "Theories, Unscientific." Such a grouping seemed to encompass many of the speculations of the new cosmology. The wormhole speculations all fell under this heading; they went well beyond the energy scales that could be tested today or in the foreseeable future. But many more respectable theories—the various GUTs, along with inflationary cosmology and superstring theory—did so too. All were utterly unsusceptible to anything that could be considered a scientific test.

The growing chasm between theory and experiment or observation seemed to be growing wider every day. Inflation, though an appealing and plausible paradigm, had yielded virtually no testable predictions. The situation on the particle side seemed even more extreme, if that was possible. Many of the finest particle theorists had become wholly preoccupied with superstrings, a set of mathematical suppositions that would not be subjected to experimental testing within the lifetimes of anybody working on them in the 1990s.

In other scientific arenas scientists were making wonderful progress: in fluid dynamics, in the new sciences of fractals and chaos, in electronic and computational technologies, in advanced materials, in medicine and biology (excluding the highly speculative theories relating to the origin of life on earth), in superconductivity, in observational astronomy and elsewhere.

Yet cosmology seemed to be becoming less scientific. As Sheldon Glashow had suggested, it was becoming more like a medieval theology every day.

Physicists were now using their minds and their mathematics to explore regions of the universe from which humanity had once seemed eternally barred. The implications were profound; metaphysical questions and even religious issues would surface in these directions. Yet the answers many physicists were giving to fundamental questions—Where did the universe come from? Why does it contain matter?—were being labeled science. In fact, these answers could be considered at best only informed speculation.

The so-called top-down approach to dissecting nature was indeed tantalizing. How delightful to construct answers from pure thought without resorting to the scientific elbow grease of long and hard experimentation. How bold to explain everything in the universe in a single aesthetically and intuitively satisfying mathematical stroke based on arguments from design.

However, such an approach had not done well in the past, with a priori answers failing ever since the time of ancient Greece. Without astronomical observation, we would still be inhabitants of a Ptolemaic universe. Without the provocation of Galileo's experiments, Newton would not have divined gravity. Without the failure of the Michelson-Morley ether experiment, how far would Einstein have gotten on pure thought alone?

Physics, as it related to the most fundamental issues of matter and energy and the universe at large, had gone full circle and was becoming metaphysics all over again. A new science confronting its first crisis, cosmology already had passed through this veil between physics and metaphysics.

Was it now metacosmology? It was starting to look as if a few of its most respected practitioners fully intended to turn their science into a new branch of philosophy.

CHAPTER **SIXTEEN**

THE
ACCIDENTAL
UNIVERSE

*The field cannot well be seen
from within the field.*
—RALPH WALDO EMERSON

*Perhaps the first rule
I must impose on myself is this:
Stick to what I know.*
—ITALO CALVIN

During
the late 1980s Stephen Hawking
traveled around the United States and Europe giving
a lecture he called "The Origin of the Universe." Other theorists had
given similarly named talks. However, it usually turned out that they
planned to talk about something that might have happened a billionth or
a trillionth of a second after the big bang, not the actual origin itself.

Hawking intended to talk about the very beginning, the moment
that time began, the actual creation event. He would explain the very
initial conditions of the universe. He believed that if these were known,
then everything else in the history of the universe would follow according
to physical law as laid down by physicists and physicists alone. There
was no room for God in Hawking's calculations.

Only a very few theorists had the wit and gall to carry off such a
talk. Hawking had developed a new approach to the problem of the universe. He was unhappy with the theories aimed at quantizing gravity.
His own pet theory of supergravity had fallen out of vogue and he had
never jumped on the superstring bandwagon. Moreover, none of the new
theories seemed to explain what physicists called the boundary conditions of the universe, the situation extant at the very beginning and the
very end of time, the ultimate parameters.

Hawking had come up with a no-boundary idea that he believed

211

was the key to the beginning conditions of the universe. If you considered that the singularity at the big bang, which he and Penrose had mathematically demonstrated twenty years earlier, was the initial boundary, then he believed you just might be able to extrapolate backwards in time a few more ticks of the cosmic clock to time zero itself.

If not, then what was the good of modern theoretical physics? Your theory of the universe would not be complete. You had to know the initial condition. Otherwise, you were reduced to merely describing the universe: "It is as it is because it is," was, he thought, not very satisfying.

Hawking believed that as you went back in time, the universe really did display increasing symmetry. The electroweak unification, and its experimental confirmation in Carlo Rubbia's big accelerator, seemed to show that. Even if the GUTs couldn't yet be proven, nor the superstrings, there was no reason to think that the symmetry of the universe at an earlier time would be any less.

The farther back you went the smoother the universe would become, and as you passed through the supposedly impenetrable Planck time at 10^{-43} second after the big bang, the minuscule embryo of a universe would be quintessentially smooth and symmetric. Hawking worked with James Hartle, the Santa Barbara theorist, on the metaphysical problem of the initial condition.

To begin with, they asked what effect quantum mechanics would have had on highly compacted space-time. As they drifted back on the wings of the geometry of general relativity and the equations of quantum mechanics, they found that the uncertainties discovered by Heisenberg would play havoc with an extremely dense infant cosmos, actually eliminating any distinguishing features of space and time.

Here time would become simply another dimension of space, or "spatialized" as Hawking put it, with the universe existing only in four dimensions of space. Imaginary numbers were essential to these calculations.* If these were applied to the concept of time, then time would lose its essential timelike character of always pointing in one direction—the so-called arrow of time. This imaginary time would point in either direction; in essence it would be no time at all since it would have ceased to exist.

This meant that there had been nothing before the beginning. Nor, for that matter, would there be anything after the end if the same reasoning were applied to the final contraction of the universe in the great gravitational day of atonement that physicists liked to call the big

---

*Imaginary numbers were invented by Descartes, Newton, Euler and others to define the square of any negative number. There is no real, or ordinary, number that when multiplied by itself will produce a negative number.

crunch. In the absence of time, *before* and *after* were expressions that were simply irrelevant.

If you were to ask Hawking what happened before the big bang, he usually responded that this was like "asking for a point one mile north of the North Pole." In the Hawking-Hartle cosmology, the four-dimensional space at the beginning (of course, we couldn't even use that expression) would simply curve back on itself, forming a little closed dimensionless spherelike surface. To demonstrate this fact at his lectures, he usually had an assistant draw a four-dimensional closed sphere on the blackboard. This was intended to show that there was no beginning and that there would be no end. The universe was simply without boundaries.

In between these non-boundaries at the beginning of expansion and the end of contraction, of course, the normal laws of nature would prevail, running the universe in orderly fashion. Hawking and Hartle had eliminated the need for a god at all. Physics could explain everything from one non-boundary to the other, as well as everything in between.

Why is there something rather than nothing? Hawking and Hartle had a simple answer. The universe started out as close to nothing as you could imagine. And to fully imagine this nothingness you had to resort to the mathematics of imaginary numbers and imaginary time.*

The Holistic Universe

Hawking said often that he was after nothing less than a complete geometric picture of the entire history of the universe. He wanted to know its structure in space and in time. Anything less would not do. He wanted the entire universe in a set of calculations.

Even he admitted that it was an arrogant goal. Perhaps the goal was the result of his own inner fears that his physical condition might not allow him time for attaining anything less than the big picture. In terms of envisioning the universe as a unified system in both time and space, Hawking was John Archibald Wheeler's greatest protégé in a sort of mathematical-spiritual sense, although the two men were never especially close personally or socially.[1]

Wheeler had been around the block as far as twentieth-century physics was concerned. He was a symbolic link between the great revolutionary discoveries at its beginning and the uncertainties at its close. While a graduate student at Princeton he had fallen into Einstein's general relativity circle, then later studied quantum mechanics under Niels

---

*In his 1988 book *A Brief History of Time,* Hawking did not distinguish between *imaginary* as it related to mathematics and the way it is used in everyday language, something that may have confused nontechnical readers.

Bohr in Copenhagen. Of the two men, Wheeler considered Bohr the greater human being.

"You can talk about people like Buddha, Jesus, Moses, Confucius, but the thing that really convinced me that such people existed were conversations with Bohr," Wheeler once said. "I would regard Einstein as declaring the goal we should work toward and Bohr as providing the method and style." [2]

In 1939 when Bohr visited the United States, Wheeler met him at the pier in New York. Bohr pulled him aside and whispered the awful news that German scientists apparently had been able to split uranium atoms. The two soon produced a paper that laid out the theoretical groundwork for the atomic bomb. Wheeler collaborated with Bohr on one project or another for years. When Bohr died in 1962, Wheeler broke down when he made the announcement to a class he was teaching at Princeton.

Wheeler's mind was among the most inventive of any theorist I ever met. He was famous for coming up with radical new ideas—and just as famous for giving them away. "His ideas are always strange," Richard Feynman, Wheeler's former student at Princeton, said once when I visited him at Caltech. "I don't believe them at all. But it is surprising how often we realize later that he was right."

In field after field—astronomy, quantum mechanics, nuclear physics, cosmology—his ideas often catalyzed major discoveries. When Feynman accepted his Nobel Prize, he told the audience that it was Wheeler who had given him a main feature of his theory during a telephone call. [3]

Despite his closeness to Bohr, Wheeler had always extended himself beyond his mentor when it came to speculating about the nature of the universe. Following in Einstein's footsteps instead, he had taken up the quest for the big picture. As a young physicist, Wheeler was certain that the entire universe would one day be explained in the physics of curved space-time. As he grew older and such a solution seemed forever beyond reach, he began turning to the observer-participant notions of quantum theory for an answer. He believed in his heart that physicists could achieve the answer.

"When it comes," he said one day in 1985 over lunch in the faculty restaurant at the University of Texas, "it will be so simple, so pure and so obvious that we will marvel at ourselves for not having seen it all along."

He was certain that the presence of humans in the universe went a long way toward explaining not just how the universe had come into being, but also why. This had led Wheeler to join in the rush for an idea that purported an answer to such questions. Called the anthropic principle, it came in two versions: the weak and the strong.

The roots of the weak version were in a calculation of Dirac's in

1938 showing that certain fundamental constants such as the ratio of electron to proton masses could be mathematically maneuvered to roughly equal the age of the universe. According to the idea as it has since developed, current conditions in the universe—the stars, galaxies and planets, along with the strengths of the four known forces—all came together to allow for the existence of humans.[4]

The strong anthropic principle held that _only_ in a universe with the physical constants we observed—the speed of light, the Planck constant, gravity, and so on—could human life have emerged. The universe has such constants in order to permit human existences. Most theorists recognize the strong principle for what it is: a quasi-religious top-down argument that the universe is the product of a designer-creator for the benefit of humanity. Hawking, who had a hand in developing the generic anthropic idea, later called it a "counsel of despair," while Guth said it was for people who had nothing better to do.

The anthropic idea was closely bound up with the age-old teleology of a whole-universe theory, an eternal pitfall of Western science caught in the vise between the gut hope for a final solution and the reality of observation and experiment. If you were looking for the secret of the universe, you needed a guiding principle, and the anthropic idea in one guise or another seemed to provide it. This caused a great deal of confusion.

The late Heinz Pagels deplored the anthropic idea, calling it "much ado about nothing,"[5] yet in his book about quantum mechanics, _The Cosmic Code,_ he asked the question, What is the universe? His answer—"a message written in code, a cosmic code, and the scientist's job is to decipher that code"—was another version of the anthropic universe.

The anthropic idea has been around for a long time, although not in physics circles until the 1930s. In _Man's Place in the Universe_ (1903), Alfred Russel Wallace, a colleague of Darwin's in the discovery of natural selection, had proposed that had the Earth and solar system been constructed any differently, conscious life in the universe would have been impossible.

In an irreverent essay, "The Damned Human Race," Mark Twain wonderfully dispatched Wallace's anthropic ideas: "Man has been here 32,000 years. That it took a hundred million years to prepare the world for him is proof that that is what it was done for. I suppose it is. I dunno. If the Eiffel Tower were now representing the world's age, the skin of paint on the pinnacle-knob at its summit would represent man's share of that age; and anybody would perceive that that skin was what the tower was built for. I reckon they would. I dunno."

Wheeler had toyed around with the anthropic universe. He recognized its failings. Still, he was intrigued with the idea of a participatory universe. He looked at the uncertainty principle. This was one of the strangest things in the universe, he thought. It made him believe that what we can say about the universe as a whole depends on the means we use to observe it. If to measure a particle is to decide which of its properties has tangible reality, then a physicist is not simply an observer, but also a participant. Human beings, by exploring the universe, played a part in bringing into being something of what they saw.

"Laws of physics relate to man, the observer, more closely than anyone has thought before," Wheeler said when I visited him in Austin in 1985. "The universe is not out there somewhere independent of us. Simply put: without an observer, there are no laws of physics."

Wheeler asked himself if all the billions of acts of observer participation add up to everything that we call the universe. His answer: "Nobody has ever proposed another way of bringing into being plan without plan, natural law without law, matter without matter." [6]

Wheeler's speculations were always the most fun of anybody's, partly because as an elder statesman of theoretical physics he no longer worried about what anybody thought and partly because he had a playful, jubilant mind. He always wanted to be at the frontier of current intellectual ideas, or, he hoped, beyond it. He believed it was his duty to investigate the strangest things about the universe. He was full of dictums.

"In any field, whether it's psychology or physics, find the strangest thing, and then explore it," was one of his favorites, and he followed it to a T.

In attempting to explain why the universe existed, Wheeler explored the very strange and very elusive edges of the universe. Ironically, he sided against his mentor Bohr, whose long intellectual struggle with Einstein during the middle decades of the twentieth century had helped define humanity's description of the universe.

Einstein's attack on quantum theory was in reality a battle for the very soul of physics. In all likelihood, he could have lived with quantum theory's counterintuitive, Lewis Carroll-like treatment of a particle's position and momentum. What he could not abide was that quantum mechanics ruled out determinism in the submicroscopic world. He had a passionate, deeply held belief in the objective reality of the universe, in the laws of cause and effect and, most especially, in the capacity of the human mind to truly understand what he called "the Old Man's secret."

Bohr agreed with Einstein that outside the atomic realm standard logic would generally apply. But within the world of the atom, Bohr said, "Evidence obtained under different experimental conditions cannot be comprehended within a single picture." [7] He called this philosophical viewpoint complementarity. It was an insight that had come to him following a skiing holiday in 1927, and it was a revolutionary concept.

In 1915 Bohr's friend, the Danish psychologist Edgar Rubin, wrote a paper on concepts of visual perception similar to those then being developed in the new gestalt school of psychology. Using classic figure-ground diagrams, Rubin pointed out that it was virtually impossible for anybody to see both pictures in one of these visual illusions at the same time. Psychologists interpreted this as a reflection of a basic limitation of the brain's capacity to process visual information.

Thomas Kuhn had used the gestalt diagrams as a metaphor for competing scientific paradigms. Similarly, Bohr interpreted the diagrams as analogous to the mind's relationship to quantum mechanics. The paradoxes of quantum theory reflected a basic limit on the mind's ability to conceptualize reality. With the gestalt illusions, a slight shift of perspective could refocus the mind's eye from one figure to another. In quantum physics, Bohr believed, a shift in the physicist's experimental equipment brought about a similar change in perception: In one case, an electron would be seen as a particle, in the next, a wavelike entity.

Scientists, even Einstein, had to accept these limitations on perception, Bohr insisted. Quantum theory was simply the best that the human mind could accomplish as it explored the atomic realm. There it would always encounter a permanent barrier to additional understanding in the form of something similar to a gestalt diagram of competing perceptions.

Einstein and Bohr danced through a kind of gestalt of physics. Their two approaches were simply incompatible. If you took Einstein's viewpoint, you could not take Bohr's. Both ideas, of course, had a long and noble history. General relativity and its offshoot, continuous field theory, was but a modern instance of an age-old holistic world view, while Bohr's reflected the atomism of Leucippus and Democritus, a belief in a system of separable things that could only be described separately and distinctly and in terms of probabilities.

The intervening decades seem to have tipped the scale of scientific balance toward Bohr's view, ruling out the possibility of performing certain measurements capable of defining fundamental concepts about nature. Even after all the decades of hard, driven work by Einstein and many, many others, quantum mechanics and general relativity were as incompatible as ever by the early 1990s. With the developing crisis in cosmology and the experimental failures and proliferation of untestable theories in particle physics, was there any reason still to believe in the universe as a holistic system?

Was There a Big Bang?

A 1985 novel by Italo Calvino provides a wonderfully instructive little analogy to what was going on in cosmology in the late 1980s and 1990s. The title character, Mr. Palomar, constructed models in his mind to explain the world he saw around him. This was the only way Mr. Palomar believed he could understand the most entangled human problems involving individual relationships, society or the government. He followed the same deductive procedures developed by physicists and astronomers who investigated the structure of the universe and of matter.

First, he constructed a mental model that was the most perfect, logical, geometrical model possible; then he looked around to see if the model was suited to the observations of practical experience; then he tried to make the corrections necessary for model and reality to coincide. He believed that the construction of a model was a miracle of equilibrium between shadowy principles and elusive experience, with the result more substantial than either.

He attempted to fit every experience of his life into his models and ultimately into a model of models. There was a model to help him see how to weed the lawn, how to buy wine, how to select his clothes for the day, how to behave in society, how to look at the universe, whether as "regular and ordered chaos or as chaotic proliferation." He believed that every detail of a well-made model must hold together in absolute coherence with every other detail.

"A model is by definition that in which nothing has to be changed, that which works perfectly, whereas reality, as we see clearly, does not work and constantly falls to pieces," he thought. "So we must force it, more or less roughly, to assume the form of the model."

For a long time Mr. Palomar believed that all the difficulties, inconsistencies and contortions of human reality and the greater universe beyond that failed to fit his model were but transitory, irrelevant accidents. After all, what counted was only the serene harmony of the lines of the pattern. For Mr. Palomar, the models he imagined eventually became "a kind of fortress, whose thick walls conceal what is outside."

He sought revelation from his models, but he found he was always overwhelmed by a flood of experiences. Gradually Mr. Palomar finally became convinced that what really counted was what happened despite the models.[8]

Cosmologists constructed models to explain the structure of the universe. Often, as we've seen over and over, they resort to extreme measures when their models do not accommodate newly discovered factual details. They have done this in two ways: either by constructing models with arbitrarily adjustable parameters, the case in many of the computerized projections of the universe based on cold dark matter; or by creating a number of slightly different models, the case with the various inflationary scenarios or the standard model of particle physics. In both cases, every bet is hedged with suspicious flexibility.

Ptolemy, the ancient Alexandrian astronomer-mathematician, created a model in which the sun, moon, planets and stars traveled in perfect circles around the Earth at the center of the universe. As time went by, ever more accurate observations showed that planetary orbits were not circular, and a new theory of planetary motion was introduced that involved adding complex epicycles to account for the extraneous motion.

As the centuries passed, astronomers varied the sizes of the epicycles or piled theoretical epicycle upon epicycle to match astronomical observation. By the sixteenth century the calculations needed to make the model fit observational details were too cumbersome for all but the most mathematically adept.

Today, in a trend reminiscent of the methods of Ptolemaic astronomers until Copernicus came to the rescue, cosmologists have begun playing fast and loose with facts that fail to fit their models. The big bang theory, the reigning cosmological model of our time, is a case in point. When the theoretical model first appeared, it was a reasonable and seemingly scientific explanation for a relatively small amount of astronomical data taken earlier in the century.

For one thing, it was consistent with the Hubble redshift of galaxies and large-scale expansion; for another, it seemed to explain the observed abundances of light elements such as helium and hydrogen. Helium and hydrogen had not been created in the fusion furnaces of existing stars, physicists believed, but forged in the fiery crucible of the earliest moments of the big bang.

In what probably was its finest hour, the model predicted the microwave background radiation at about the temperature that was consistent with a creation explosion out of a formless nothingness 15 billion years or so ago. By the end of the 1960s, with Fred Hoyle's competing steady-state theory apparently laid to rest, the big bang appeared to be such a simple and elegant and true explanation that it was accepted by virtually everybody.

Two decades later it was so much a part of both the popular consciousness and the physics community that it was easy to forget that the big bang was nothing more than a theoretical model. However, despite everybody's enthusiasm for it, the troubling observational and theoretical problems of the 1970s and 1980s increasingly challenged the big bang model's most fundamental explanatory authority.

Another problem is that scientists have viewed the big bang as an event without a cause. Being humans, physicists would, of course, always be curious about the absolute origin of the universe. Yet Aristotle, in the fourth century B.C., had pointed out there could be no such thing as a first event. If a single occurrence such as the big bang had happened, one always had to ask, "Why then, why not earlier?"

The only answer could be, "Conditions were not yet right." What did it mean for conditions to be right or not to be right? There always had to be an earlier event. Hawking had proposed an ingenious solution to the conundrum by sidestepping the boundary problem in simply declaring that no boundary existed. Yet the no-boundary proposition was little more than interesting speculation that would have attracted little notice had it come from anyone less than the world's most famous physicist.

The big bang model also began having more and more difficulty reconciling the latest observational details found by astronomers with the fundamental assumption that on the galactic and cosmic scales gravity was the major player. In other words, the theory failed to explain convincingly how matter had become organized in clusters of galaxies and superclusters. For the universe to be structured in a manner consistent with current observations, more than 90 percent of its matter would have to be in the form of some strange, unknown, unseen, mysterious but unbelievably massive dark matter.

Not only would this ad hoc dark matter have to be present in such a huge quantity that it would account gravitationally for the size and behavior of the new clusters and superclusters, but it also would have to be of such a bizarre quality that it could not possibly be detected by even the most sophisticated technology. Even if it were detectable, the mythical dark form of matter would still be unable to deal with the immense superclusters discovered in the late 1980s.

For instance, gravity working alone would have taken something like 100 billion years to create the supercluster two and a half billion light years across that was recently discovered by American and German observers. This was a time scale at least five times longer than permitted by even the most generous of the big bang models.

Moreover, evidence for the big bang as well as many other cosmological theories remained intangible as to both time and space. Some cosmologists argued that the big bang theory was analogous to the theory that the world is round. Yet we *know* the earth is a sphere: We and it occupy the same time period and the same location in space, neither of which is true of the big bang. Cosmologists also have pointed to the fossil record supporting the theory of evolution as another analogy to the big bang. Yet fossils provide tangible evidence, while most of the evidence for or against the big bang is millions of light years away and not subject to direct examination. The event itself, if it occurred, is lost billions of years in the past, conveniently out of sight if not out of mind, and it is not expected to give an encore performance soon.

To salvage the big bang, theorists have brought in a number of ad hoc assumptions such as inflation, the little burp at the dawn of time when the universe was supposed to have expanded exponentially. Unfortunately, the various inflationary models are based in one form or another on one or more of the grand unified theories that purport to bring together electromagnetism, the weak nuclear force and the strong nuclear force.

Both inflation and the GUTs suffer from the same malady, an inability to make predictions that can be tested in a meaningful manner.

In the one instance in which a GUT did make a prediction, that of proton decay, the theory failed the test completely. Yet the theorists seemed absolutely determined to hang in there, come hell or high water or another observed fact inconsistent with their theoretical models.

The so-called standard model that particle physicists use to explain the relationship of the elementary entities of matter with three of the four known natural forces has been somewhat more successful than inflation or the GUTs in making at least a few predictions. It did, for instance, predict the new quarks as well as the W and Z particles discovered at CERN by Carlo Rubbia.

Yet the Higgs particle has remained elusive in spite of a new generation of particle accelerating machines. Much farther out on the slippery limb of unpredictability sit the theories of everything and the wormhole and many universe speculations of Stephen Hawking, Sidney Coleman, Kip Thorne and the others in the metacosmological business.

When he began recognizing the fallibility of his models in the face of the tangle of facts he observed in the world around him, Mr. Palomar felt the only thing he could do was erase from his mind all models and models of models. Having taken this step, he came face to face with reality, "hard to master and impossible to make uniform."

Unfortunately, most cosmologists have yet to see the light discerned by Palomar. They seem determined to hold onto the big bang model as their preferred cosmology in spite of the ever greater need to bandage its gaping wounds with ad hoc assumptions. Thomas Kuhn, it seems, was right: It is all but impossible for a scientist to let go of a long-standing paradigm when one's career has been devoted to its study—even when the paradigm no longer works.

A few alternative ideas for the formation and structure of the universe have begun springing up. None is especially compelling or necessarily more credible than the big bang. Two of them are, however, worth mentioning for their alternative treatment of the main evidence brought forward for big bang cosmogenesis. One proposed by Hannes Alfvén, a Swedish Nobel laureate, suggests that the universe has always existed, at least as far as humanity's perspective is concerned, and has been shaped as much by electromagnetic currents as by gravity.

In Alfvén's cosmology, the uniform microwave background might be a sort of radiation fog whispering throughout the universe. It could be the result, Alfvén's followers believe, of the continuous emission and absorption of electrons by the strong magnetic fields of filaments jetting out from the billions of galaxies. "Suppose you're out camping and you wake up in the morning and see an even white glow all around you," one of Alfvén's adherents observed. "You don't assume you're looking at the beginning of the universe. You just realize that a fog has come in during the night."[9]

The irrepressible Fred Hoyle, in the meantime, has rebounded with a new version of the steady-state universe. In an article published in the British science journal *Nature* in 1990, Hoyle and four other scientists questioned the conventional analysis of the redshifts of quasars and galaxies.[10] If the universe is expanding, a higher redshift supposedly indicates a greater distance from the Earth. One of Hoyle's collaborators, Halton C. Arp of the Max Planck Institute for Astrophysics in Garching, Germany, has pointed out that a number of celestial objects have been found since 1970 with redshifts that do not appear to correlate with distance from Earth.

"One point at which our [big bang] magicians attempt their sleight-of-hand is when they slide quickly from the Hubble redshift-distance relation to redshift velocity of expansion," Arp observed in 1991. "There are now five or six whole classes of objects that violate this absolutely basic assumption. It really gives away the game to realize how observations of these crucial objects have been banned from the telescope and how their discussion has met with desperate attempts at suppression." [11]

Several quasars have redshifts so great that it would appear that they are on the very edge of the universe. According to Arp, some of these quasars have been found in the vicinity of nearby galaxies with small redshifts. If the quasar were connected with the galaxy, then the two objects obviously would not be moving at vastly different velocities. This would mean, Arp thinks, that their redshifts—perhaps even all redshifts—result from a phenomenon other than rapid recession.

Redshifts also might result from the contraction of certain celestial objects. Arp believes that the quasars with high redshifts may actually be shrinking rather than receding. Another possibility is that certain kinds of light shift toward the red end of the spectrum as they propagate through space. This mechanism was suggested by Emil Wolf of the University of Rochester in 1987 and later verified experimentally.

"When big bang proponents make assertions such as 'an expanding universe . . . very well verified observation,' 'a whole bunch of observations that hang together' and 'the evidence taken together . . . hangs together beautifully,' they overlook observational facts that have been piling up for 25 years," said Arp.

"Of course, if one ignores contradictory observations, one can claim to have an 'elegant' or 'robust' theory. But it isn't science." [12]

Did the big bang happen? Maybe the big bang was just a big bang, an explosion in our little neighborhood of the universe that was neither the beginning of time nor the creation of the cosmos. Nobody knows. As humans, we are all cosmologists. Physicists and nonphysicists alike, we are all members of a lonely species staring at eternity from a small corner of the universe, driven by our intelligence and curiosity to learn our origins for ourselves.

In wondering about these origins, we seem to have found it necessary not only to seek the raison d'être for the cosmos of which we are such a minuscule aspect, but also to cast our own psychological need for understanding out into a universe that we have at least partly created in our search for epiphany. That from earliest times we have believed ourselves capable of finding the answer to the question of origin is, I suppose, the hubris of every age.

It is remotely possible that the professional cosmologists have gotten it right. Yet the evidence for the big bang is sketchy at best. In the context of new and contradictory observations, the big bang theory begins to appear more and more like an overly simplistic model in search of a creation event. By the early 1990s the big bang model was at best the pinnacle of a chain of inference that was increasingly unable to answer the most fundamental questions.

How long will the big bang theory survive? Until the end of the century? For twenty years? For fifty? More than a few theorists have expressed the opinion that it would not even last out the 1990s.[13] Whether the big bang goes down in five years or twenty-five years, it appears inevitable that it soon will be overwhelmed by more and more uncompromising new observations and experimentation. In the next millennium scientists and other people looking back likely will regard it much the way we look back on the cosmology of Aristotle, a quaint little backwater theory that people believed in for a while.

For the time being, the big bang remains a scientific paradigm wrapped inside a metaphor for biblical genesis, a compelling although simplistic pseudoscientific creation myth embodying a Judeo-Christian tradition of linear time that led to Western ideas about cultural and scientific progress and which ordained an absolute beginning.

An Evolutionary Twitch

Adam and Eve were banished from Eden for eating fruit from the tree of forbidden knowledge. In _Paradise Lost_ (1667) Milton records a mythical conversation in which Adam, before Eve's fatal trespass, asks the angel Raphael to tell him the secret of creation.

"From Man or Angel the great Architect did wisely to conceal, and not divulge his secrets to be scann'd by them who ought rather admire," replies the angel.

Doctor Faustus sought secret knowledge from dark authorities, then lost his soul. In Mary Wollstonecraft Shelley's novel _Frankenstein_ (1818), Frankenstein used a forbidden art to bring inanimate matter to life, and then is pursued by the monster he has created. Modern horror movies picked up the theme. A penalty must be paid for crossing certain natural boundaries.

Flowing through human history has been an undercurrent that certain knowledge was eternally forbidden, forever past the limits of mere mortals. Was this true for knowledge about the universe at large? Newton, caught between his science and his religion, seemed to think there was a secret of the universe, but thought it probably was beyond the power of humans to divine.

"As a blind man has no idea of colors, so we have no idea of the manner by which the all-wise God perceives and understands all things," Newton wrote.[14] Yet "the Old Man's secret" beckoned, and science kept pushing aside veil after shadowy veil in the hope and expectation that the forbidden answer would lie just behind the next curtain.

Why do we still believe that the universe holds a single great secret about to be revealed? More than ever, there was every reason to think that humanity's perception of the universe was bound up in an extension of Bohr's idea of complementarity. There was not just one secret. There were many, many, secrets.

The void between quantum mechanics and general relativity was as great as ever. Moreover, the structure of the universe as it was being filled in by observational astronomers was yielding not a grand secret, only a confusing welter of indecipherable patterns that left theorists grasping to catch up with ever more complicated and unlikely theoretical schemes.

On the particle side, it was believed that the big accelerators were within a trillionth of a second of truth and counting. Yet that parameter so tantalizingly close to the beginning of time had been established from within the closed system of the cosmos without benefit of the full picture. During the 1930s Kurt Gödel, the remarkable Czechoslovakian-born Austrian mathematician, demonstrated in his incompleteness theorems that the proof for the validation of any system could not be established from within that system.

There must be something outside the theoretical framework—whether the framework was mathematical, verbal or visual—against which a confirming or disconfirming test could be made. You could not measure the outside dimensions of a house from inside the kitchen. In other words, Gödel's theorem suggested, no theory of the structure of the universe could be validated from within that structure.

Nor, by extension, could such a theory be invalidated since by definition it would be nothing more than a metaphysical statement of conjectural assumption. Any theory by its very nature required for its validation a larger reference frame. If such a reference frame existed, then the theory could not be considered comprehensive.[15]

Beyond the particle physicists' first trillionth of a second lay a desert realm of time and energy, a terra incognita that would indefinitely remain beyond the capabilities of even the greatest of accelerators to explore. As more than one physicist has expressed it, "God has always seemed to be just around the corner." The first minute or the first second or the first split second would persist as yet more corners were turned. But it is unlikely physicists will ever turn the last corner. There always would be another one just ahead, meaning that the trillionth of a second that particle physicists believe lies between them and revelation may just as well be a trillion years.

Can we ever hope to fully understand the universe? Jim Peebles said he thought the only clean part of current cosmological theory was that the universe was expanding. I think even that modest statement needs to be modified. We know only that the universe is changing. Perhaps cosmologists should rethink their teleological approach to the universe and adopt the viewpoint of biological scientists studying evolution. The universe, like natural selection, is not only wonderfully indifferent but has not the merest afterthought about the fate of itself, much less about the fate of the human race.

What do we see when we look out there? Probably but a small corner of a vast evolutionary process of which we are only remotely aware, a tiny and accidental twitch of something incalculably more immense falling within our view: a universe perhaps finite yet probably not computable, possibly unstable within its boundaries and maybe capable of producing a multitude of other universes inside and outside itself, a great cosmic drama in an untold number of acts, a vast firmament of quasars and quarks, nebulae, dust, galaxies and galactic clusters, dark matter and light matter and shadow matter, force fields and intersecting force fields, all with or without beginning, yet perhaps without creation . . .

Here a word, there a word.
Pretty soon you have a whole book.
—JOHN BROCKMAN

absolute zero The temperature, –273° Celsius, at which all molecular activity halts.

absolute space Space that exists as an independent cosmic framework without reference to its contents, also called Newtonian space.

absolute time Linear time that exists without reference to events or spatial dimension, also called Newtonian time.

acausal initial conditions A seeming paradox in which the initial conditions of the universe are thought by some cosmologists to have occurred without prior physical conditions.

accelerator A machine, usually quite large, which is used to accelerate sub-atomic particles to great velocity. By observing the collisions of these particles with other particles, physicists are able to learn about the nature of the particles. The machines are popularly called atom smashers.

anthropic principle An effort at deducing basic information about the universe from the fact that human life exists on Earth. This principle states that the universe must have certain properties if it supports intelligent life capable of perceiving it.

antimatter Some cosmological theories maintain that for every particle in the universe there exists an antiparticle with identical spin and mass but with an opposite charge. Little antimatter has been observed in nature, but it can easily be produced in accelerators.

atom The basic unit of a chemical element, made up of electrons around a nucleus which, in most cases, consists of protons and neutrons.

baby universe According to highly conjectural theories of Stephen Hawking and others, a baby universe is created during the reproduction of one universe by another.

baryons A class of heavy particles including protons and neutrons that are subject to the strong nuclear force. The matter of everyday life on earth consists of baryonic matter.

big bang theory A group of theories formed by scientists who believe that the universe began in a hot, dense state, which then cooled and expanded into today's universe.

binary galaxies Two galaxies that orbit one another as a result of their mutual gravitational attraction.

blackbody radiation A kind of radiation that can be completely described in terms of its temperature, produced after a large number of particles have come into thermal equilibrium with one another.

black hole A hypothetical cosmic object with such a powerful gravitational field that its escape velocity is greater than the speed of light, believed to be the compressed remnant of a star whose nuclear furnace has burned out.

boson Elementary force-transmitting particle such as photons that transmit the electromagnetic force and the W and Z particles that carry the electroweak force.

bottom-up formation of galaxies A speculative scenario in which galaxies form first from condensed gas, with larger clusters and superclusters appearing afterward. See *top-down formation of galaxies*.

boundary conditions Physical conditions that determine the progress of a physical system according to the laws of nature. Theoretically, the evolution of every system can be determined by its initial conditions and its boundary conditions.

broken symmetry A process in which the instrinsic symmetry of a physical system is broken. According to many cosmological theories, there was greater symmetry in the early universe than in the one of today. These early symmetries were broken as the universe expanded and cooled.

bubble universe According to one early inflationary universe model, bubble universes were regions of empty space separated by walls of energy.

bubbles In the distribution of galaxies, according to recent observations, many galaxies are located on thin, spherical shells called bubbles.

charge-coupled devices Also known as CCDs, these are photoelectric devices capable of recording the intensity of tiny amounts of light from astronomical sources.

chaotic inflation Introduced by Andrei Linde, a highly conjectural version of the inflationary universe model, in which random quantum fluctuations produce new universes.

closed universe A universe with a finite size although it has no boundaries. In a closed universe, the expansion of space will expand for a finite amount of time, then begin contracting. See *flat universe* and *open universe*.

cold dark matter See *dark matter*.

complex number Discovered in the nineteenth century, complex numbers are the sum of an ordinary number and an imaginary number, which, unlike an ordinary number, produces a negative number when multiplied by itself.

cosmic background radiation A pervasive shower of radio waves coming from every direction of space. Big bang models postulate that this background radiation resulted from the collisions of particles when the universe was very young and that ended when the universe was about 1 million years old. Thus, the cosmic background radiation is believed by most physicists to be a remnant of the big bang and is frequently cited as one proof that the big bang actually occurred.

cosmological principle The belief held by most cosmologists that the universe is homogeneous throughout—that it is the same out there as it appears from down here.

cosmological constant A constant first introduced by Albert Einstein in the equations of general relativity and corresponding roughly to a repulsive force pervading the universe. The cosmological constant appears to be very close to zero, although theorists speculate that it should be very large.

cosmogony The study of the universe's origin.

curved space Space is curved, according to the theory of general relativity. Thus the geometry of the universe does not correspond to flat Euclidean geometry.

dark matter Unseen matter in the universe that has been detected by its gravitational effects. By some estimates, dark matter accounts for 90 percent of the mass of the universe. Its composition remains a mystery.

density fluctuation An inconsistency in an otherwise smooth distribution.

domain wall A two-dimensional flaw in space-time that separates the universe into different domains.

Dirac equation Describing the behavior of the electron, the Dirac equation incorporates quantum mechanics and general relativity.

electromagnetic force Fundamental interaction that acts on electrically charged particles such as the electron.

electron volt An amount of energy required to push an electron across one volt.

electroweak force A supposedly fundamental force described by the electromagnetic and weak nuclear forces.

equivalence principle In general, the principle that the force of gravity in one direction is equivalent to an acceleration in the opposite direction.

ether According to classical Newtonian physics, an invisible medium pervading space and thought to carry light waves. According to Aristotle, stars were made of ether, which was thought of as the fifth element.

event horizon The boundary around a black hole through which no matter or radiation can escape.

Faber-Jackson relation The observed correlation between the velocities of stars in a galactic core with the luminosity of the galaxy—the higher the speeds, the greater the luminosity.

false vacuum When a region of space appears to be empty, but contains stored energy, it is said to have a false vacuum.

fermion A class of particles including electrons, protons, neutrons and quarks.

flat universe At the boundary between an open and closed universe, a flat universe has an average curvature of space equal to zero. In such a universe, space is infinite and its expansion eternal.

force There are believed to be four of these in the universe, each one an interaction that brings about change in a system. The known forces are gravity, electromagnetism and the strong and weak nuclear forces.

galactic halo The unseen, nonluminous mass that is believed to surround every galaxy.

galaxy A large conglomeration of stars bound together gravitationally, typically containing about 100 billion stars.

general relativity Einstein's theory, formulated in 1915, that describes how gravity is produced by the interaction of matter and space-time.

GeV A billion (10^9) electron volts.

gluon The particle binding quarks together.

gravitation The force responsible for attracting conglomerates of matter to one another.

gravitational lensing   An effect occurring when a massive object such as a galaxy bends light to create multiple images of a more distant object.

graviton   Conjectural force-carrying particle that transmits gravitation.

gravitino   Conjectural force-carrying particle in superstring theories.

Great Attractor   An unseen mass that appears to have a powerful gravitational effect on many nearby galaxies, including the Milky Way.

GUTs   Grand unified theories, highly speculative, that attempt to combine the electroweak interaction with the strong nuclear force.

hadron   Particles subject to the strong force, divided into two classes—mesons and baryons.

Heisenberg uncertainty principle   In quantum mechanics, the principle that it is impossible to measure simultaneously the position and momentum of a particle.

Higgs field   A hypothetical energy field, predicted by grand unified theories, that gives mass to particles. There is no evidence such a field actually exists. The energy stored in a false vacuum would be the energy of a Higgs field.

Higgs particle   The conjectural particle, still undiscovered, of the Higgs field.

horizon   In cosmology, the greatest distance an observer can see since light from more distant objects has not had enough time to reach us.

horizon problem   The puzzle in the standard big bang model that widely separated regions of the universe seem to possess the same physical properties even though they were too far apart in the early universe to have exchanged these properties.

Hubble constant   The rate of the universe's expansion.

Hubble law   That the recessional velocity of galaxies is directly proportional to their distance of separation. Galaxies that follow this law precisely are part of the so-called Hubble flow. Because the universe is not homogeneous, the actual motions of many galaxies do not follow this flow.

indeterminacy principle   In quantum mechanics, the principle that the trajectory and position of any particle cannot be known with certainty.

inflationary universe theory   Speculative hypothesis, first proposed by Alan Guth, modifying the standard big bang model. Inflation suggests that the very early universe underwent a period of exponentially rapid expansion for a brief period before proceeding at the linear rate of the standard model.

initial conditions   The state of a physical system at the beginning of an interaction.

large-scale structure   The distribution of galaxies over great distance. In a perfectly homogeneous universe, there would be no large-scale structure.

lepton   One of six kinds of light particles—the electron, the muon and tauon and three kinds of neutrino.

magnetic monopole   Predicted by various grand unified theories, this is a particle that would have only one magnetic pole. None has been confirmed to exist.

meson   A particle consisting of a quark and an anti-quark.

microwave   Radio wave of short wavelength.

muon   A lepton similar to but about 200 times heavier than an electron.

neutrino A lepton with no electrical charge. Its mass, if any, is unknown.

no-boundary proposal Proposal by James Hartle and Stephen Hawking formulated within a quantum theory calculation of the early universe, the geometry of the universe's initial condition would be analogous to the surface of a sphere—i.e., without specific boundary.

omega The ratio of the universe's actual density to the critical density that would be needed to halt the universe's expansion. In an open universe omega is less than 1; in a closed universe omega is greater than 1; in a flat universe omega is exactly 1.

open universe One which is infinite in extent, its expansion eternal. Its geometry is that of an infinite curved surface.

pancake model In this model of galaxy formation, the very first structures to form from the primordial gas of the early universe were very large. These coalesced into thin sheets, which eventually fragmented into galaxies.

peculiar velocity The part of a galaxy's motion that deviates from what would be expected in the general expansion of the universe.

phase transition The sudden transition from one state to another, such as may have occurred one or more times during the evolution of the universe.

photon The force-carrying particle responsible for transmitting the electromagnetic force; a particle of light.

Planck's constant The fundamental unit of quantum mechanical effects. It can be combined with other fundamental constants such as Newton's gravitational constant and the speed of light to mark critical epochs in the history of the universe according to the big bang model. At the Planck time of 10^{-43} second, for instance, the cosmic matter density was so great that the force of gravity, extremely weak in today's universe, was as strong as the three subatomic forces.

positron The electron's antiparticle.

proton A massive particle made up of three quarks and found in the nuclei of atoms.

proton decay The eventual disintegration of protons predicted by grand unified theories; yet to be observed.

pulsar A rapidly rotating stellar remnant emitting electromagnetic radiation.

quantum chromodynamics (QCD) The quantum theory explaining the behavior of quarks and the strong nuclear force.

quantum electrodynamics (QED) The quantum theory explaining the nature of the electromagnetic force.

quantum fluctuation Continual changes in the subatomic properties of a physical system brought about the probabilities of quantum mechanics. Such fluctuations cause particles to appear out of apparent nothingness and then disappear again. Several highly conjectural theories suggest that a quantum fluctuation led to the creation of our universe.

quantum mechanics The theory, developed mainly during the 1920s and the 1930s, explaining the dual particlelike and wavelike nature of matter as well as its probabilistic character on the subatomic level. It is one of science's most successful theories of nature.

quantum gravity A projected—but as yet unrealized—theory that would combine quantum mechanics and general relativity in a single formulation.

quark One of the fundamental constituents of matter on the subatomic level, a constituent of protons, neutrons and mesons.

quasars Believed to be the luminous cores of young galaxies at the edge of the universe.

redshift A shift in color toward the red end of the spectrum, believed to be the result of the outward motion of a galaxy in the general expansion of the universe.

shadow matter Theoretical kinds of particles that would interact with ordinary particles only through the force of gravitation. There is no proof that such matter actually exists.

singularity A point of infinite density where space is infinitely curved and the equations of general relativity break down.

snakewood A material used by Baron Roland von Eötvös in his gravity experiments at the turn of the century.

space-time The three dimensions of space combined with the single dimension of time used to depict events described by general relativity.

special relativity Explains how measurements of length and time are different for different observers; formulated by Einstein in 1905.

standard model A combination of two subtheories, quantum chromodynamics and the electroweak theory. Many physicists do not believe the standard model goes far enough in explaining events in the subatomic arena even though it is quite successful.

steady-state theory A model in which the universe's density does not change over time.

strong nuclear force The fundamental interaction that binds quarks together in the atomic nucleus.

supergravity The name used for a number of unsuccessful theories that in the early 1980s attempted to explain gravity and the three subatomic forces within a single formulation.

superstring theory A highly conjectural group of theories purporting to unify all the forces of nature in a single multidimensional framework. In these theories, the fundamental unit of nature is a one-dimensional object known as a string. These theories have yet to make a prediction that can be tested observationally or in the laboratory.

supersymmetry A set of theories, also called SUSY, which speculates that there may be only one class of fundamental particles rather than two—fermions and bosons. There is no evidence that this is the case.

symmetry In physics, the property of a system such that it remains unchanged even after some kind of transformation.

tauon The heaviest of all leptons, about 3,500 times more massive than an electron.

theory of everything A theory from which every law of physics could be derived mathematically. Some physicists believe that one of the superstring theories may eventually prove to be a theory of everything, or TOE. Others think that such a theory is an impossibility.

time In physics, a dimension that separates past, present and future. In the equations of general relativity, time is viewed as geometrically analogous to spatial dimensions. In ordinary life on Earth, time is viewed as a one-way flow.

top-down formation of galaxies A hypothesis that large concentrations of matter formed first before fragmenting into individual galaxies; the reverse of the bottom-up formation of galaxies.

virtual particle According to the indeterminacy principle of quantum mechanics, a particle can pop into existence in a pure vacuum even when enough energy to create such a particle is not present. A virtual particle soon disappears in order to pay off its energy debt.

W and Z particles Force-carrying bosons that transmit the combined electromagnetic and weak nuclear forces, they were predicted by the electroweak theory of Steven Weinberg, Abdus Salam and Sheldon Glashow in the 1960s and discovered at CERN in the 1980s.

white dwarf A star, about the size of the Earth, in the final stages of gravitational collapse, which begins hot and white and ends cold and dark.

weak nuclear force Fundamental interaction that causes radioactivity and is considered a part of the more general electroweak force.

wormhole A theoretical object that would function as a bridge to another region in space or time. Even if they actually exist, they are so minuscule that it is unlikely they will ever be seen.

z The notation for an astronomical object's redshift.

All illustrations are by Wendy W. Cortesi who should be given full credit, except as noted:

"The Great Wall," chapter three, is reprinted with permission of the Smithsonian Astrophysical Observatory.

Special appreciation to Charles Misner of the University of Maryland for his expert assistance on "The Horizon Problem" in chapter four.

The "Large Electron-Positron Collider" in chapter ten, by Wendy W. Cortesi, is based upon an illustration in "The LEP Collider," by Stephen Myers and Emilio Picasso, in *Scientific American,* July 1990, p. 55.

The "Thermal History of the Universe" in chapter twelve is by Jan Adkins.

Chapter One

1. John Boslough, "The Unfettered Mind: Stephen Hawking encounters the dark edges of the universe," *Science 81,* November 1981, 66–73.

2. Dennis Overbye, 1991, p. 380.

3. Ivars Peterson, "State of the Universe: If not with a Big Bang, then what?", *Science News,* 139:232, April 13, 1991. This article contains one version of Schramm's often-quoted remark.

4. Sam Flamsteed, "Big Bang Bashing," *Discover,* June 1991, p. 23.

Chapter Two

1. For more about this fascinating site, see B. M. Lynch and L. H. Robbins, "Namoratunga: The first archeoastronomical evidence in sub-Saharan Africa," *Science,* 1978, 766–8. A people called the Borano who also care deeply about time and live across the border from Namoratung'a in the lowlands of southern Ethiopia today may have descended from the builders of the ancient observatory. See C. L. N. Ruggles, "The Borano Calendar: Some Observations," *Archaeoastronomy, Supplement to the Journal for the History of Astronomy,* 11:s35, 1987.

2. V. C. Rubin, "Differential Rotation of the Inner Metagalaxy," *Astronomical Journal,* 56:47, 1951.

3. An unusually interesting fellow, Gamow had a marvelous sense of humor and was an inveterate practical joker. In one well-known incident, which bears repeating one more time, he was about to publish a paper with a collaborator named Ralph Alpher about the creation of certain elements during the big bang. With malice aforethought, Gamow enlisted particle theorist Hans Bethe of Cornell to join him and Alpher as authors of the paper, making the author line read Alpher, Bethe, Gamow (R. A. Alpher, H. A. Bethe, G. Gamow, "The Origin of Chemical Elements, *Physical Review,* 73:803, 1948). In a stretch this could be pronounced *alpha, beta, gamma,* the first three letters of the Greek alphabet.

4. V. C. Rubin, "Fluctuations in the Space Distribution of the Galaxies," *Proceedings of the National Academy of Sciences,* 40:541, 1954.

5. V. C. Rubin, W. K. Ford, et al., "Motion of the Galaxy and the Local Group Determined from the Velocity Anisotropy of Distant Sc I Galaxies," *Astronomical Journal,* I. The Data, 81:687; II. The Analysis for the Motion, 81:719, 1976.

6. Marcia Bartusiak, "The Woman Who Spins the Stars," *Discover,* October 1990, p. 91.

7. V. C. Rubin and W. K. Ford, "Rotation of the Andromeda Nebula from a Spectroscopic Survey of Emission Regions," *Astrophysical Journal,* 159:379, 1970; and V. C. Rubin, W. K. Ford and N. Thonnard, "Extended Rotation Curves of High-Luminosity Spiral Galaxies. IV. Systematic Dynamical Properties," *Astrophysical Journal Letters,* 225:L107, 1978.

8. For a good general discussion of their discovery and its implications, see Vera C. Rubin's "Dark Matter in Spiral Galaxies" in *Scientific American*, December 1988.

Chapter Three

1. See Percival Lowell (1855–1916), *Mars and Its Canals*, New York: Macmillan Co., 1906.

2. For more on the dawn of X ray astronomy, see Riccardo Giacconi and Harvey Tananbaum, "The Einstein Observatory: New Perspectives in Astronomy," *Science*, 209, August 22, 1980.

3. S. M. Faber and R. E. Jackson, "Velocity Dispersions and Mass-to-Light Ratios for Elliptical Galaxies," *Astrophysical Journal*, 204:668, 1976.

4. For a general overview see Alan Dressler, "The Large-Scale Streaming of Galaxies," *Scientific American*, 257:38, 1987; also Somak Raychaudhury, "The Distribution of Galaxies in the Direction of the 'Great Attractor,' " *Nature*, 342, November 16, 1989.

5. R. Scaramella, et al., "A Marked Concentration of Galaxy Clusters: Is This the Origin of Large-scale Motions?" *Nature*, 338:562–564, 1989; and D. A. Allen, et al., "A Supercluster of IRAS Galaxies Behind the Great Attractor," *Nature*, 343:45–46, 1990.

6. Y. B. Zel'dovich, J. Einasto, S. F. Shandarin, "Giant Voids in the Universe," *Nature*, 300, December 2, 1982.

7. Alan Lightman and Roberta Brawer, eds., 1990, pp. 380–94.

8. John Noble Wilford, "New Surveys of the Universe Confound Theorists," *The New York Times*, January 15, 1991, p. C1.

9. V. de Lapparent, M. J. Geller, J. P. Huchra, "A Slice of the Universe," *Astrophysics Journal*, 202, L1, 1986.

10. Margaret J. Geller and John P. Huchra, "Mapping the Universe," *Science*, 246:897, November 1989.

11. J. B. S. Haldane, "Possible Worlds," in *Possible Worlds and Other Essays*, London: Chatto and Windus, 1927.

Chapter Four

1. R. Alpher, R. C. Herman, G. A. Gamow, "Thermonuclear Reactions in the Expanding Universe," *Physical Review D*, 74:1198, 1948.

2. A. A. Penzias and R. W. Wilson, "A Measurement of Excess Antenna Temperature at 4080 Mc/s," *Astrophysical Journal*, 142:419, 1965; R. H. Dicke, P. J. E. Peebles, P. G. Roll, and D. T. Wilkinson, "Cosmic Blackbody Radiation," Ibid., 142:414; and P. J. E. Peebles, "The Black-Body Radiation Content of the Universe and the Formation of Galaxies," Ibid., 142:1317.

3. Fred Hoyle, "A New Model for the Expanding Universe," *Monthly Notices of the Royal Astronomical Society*, 108:372, 1948.

4. Hoyle, 1965.

5. A. Einstein, "Cosmological Considerations on the General Theory of Relativity," *Sitzungsberichte der Preussischen Akad. d. Wiss.,* 142–152, 1917 (translated).

6. In 1989 Hoyle published a paper called "The Steady-State Theory Revived?" in *Comments on Astrophysics,* 13:81.

7. For a good general description of the instruments aboard the COBE satellite, see Samuel Gulkis, Philip M. Lubin, Stephan S. Meyer and Robert F. Silverberg, "The Cosmic Background Explorer," *Scientific American,* January 1990.

8. Author interview with Richard Isaacman, 1991.

9. Thomas J. Broadhurst, Richard S. Ellis, David C. Koo, A. S. Szalay, "Large-Scale Distribution of Galaxies at the Galactic Poles," *Nature,* 343, February 22, 1990.

10. C. W. Misner, "The Mixmaster Universe," *Physical Review Letters,* 22:1071, 1969.

11. A full explanation of the horizon problem is in Charles W. Misner's "Cosmology and Theology," 1977.

12. M. Schmidt, "3C 273: A Star-like Object with Large Redshift," *Nature,* 197:1040, 1963.

13. Donald P. Schneider, Maarten Schmidt, James E. Gunn, "A study of ten quasars with redshifts greater than four," *The Astronomical Journal,* Vol. 98, No. 6, November 1989.

Chapter Five

1. Robert H. Dicke, *Gravitation and the Universe: The Jayne Lectures for 1969,* American Philosophical Society, 1969.

2. C. B. Collins and S. W. Hawking, "Why is the Universe Isotropic?" *Astrophysical Journal,* 180:317, 1973; and Brandon Carter, "Large Number Coincidences and the Anthropic Principle in Cosmology," *Confrontations of Cosmological Theories with Observational Data,* IAU Symposium 63, 1974.

3. Alan Lightman and Roberta Brawer, eds., 1990, p. 392.

4. R. H. Dicke and P. J. E. Peebles, "The Big Bang Cosmology—Enigmas and Nostrums," in S. W. Hawking and W. Israel, eds., *General Relativity: An Einstein Centenary Survey,* Cambridge: Cambridge University Press, 1979, p. 507.

Chapter Six

1. Alan Guth, "Inflationary Universe: A Possible Solution to the Horizon and Flatness Problems," *Physical Review D,* 23:347, 1981.

2. A. D. Linde, "A New Inflationary Scenario: A Possible Solution of the Horizon, Flatness, Homogeneity, Isotropy, and Primordial Monopole Problems," *Physics Letters B,* 108:389, 1982.

3. Alan H. Guth and Paul J. Steinhard, "The Inflationary Universe," *Scientific American,* May 1984.

4. Lawrence M. Krauss, "Dark Matter in the Universe," *Scientific American,* December 1986.

5. Kim Griest and Joseph Silk, "No More Neutrino Cold Dark Matter," *Nature,* 343, January 4, 1990.

6. W. Saunders, et al., "The Density Field of the Local Universe," *Nature,* 349:32, January 3, 1991.

7. John Horgan, "Universal Truths," *Scientific American,* October 1990, p. 114.

8. J. E. Peebles, *Science,* 235:372, 1987.

Chapter Seven

1. "British astronomers at Cambridge University Observatory detect radio signals that may come from neutron stars; believe regularity of signals' source is from pulsating stars," *The New York Times,* March 3, 1968, p. 39.

2. For an entertaining look at Sandage and the problem of large-scale structure, see Dennis Overbye, 1991.

3. Lightman and Brawer, p. 283.

4. Author interview with Jeremiah Ostriker, Princeton, 1982.

5. Author interview with James Peebles, Princeton, 1982.

6. J.P. Ostriker and P. J. E. Peebles, "A Numerical Study of the Stability of Flattened Galaxies: Or, Can Cold Galaxies Survive?" *Astrophysical Journal,* 186:467, 1973.

7. J. P. Ostriker, P. J. E. Peebles, and A. Yahil, "The Size and Mass of Galaxies and the Mass of the Universe," *Astrophysical Journal Letters,* 193:L1, 1974.

8. G. Lemaître, "A Homogeneous Universe of Constant Mass and Increasing Radius Accounting for the Radial Velocity of Extra-Galactic Nebulae," *Annals of the Scientific Society of Brussels,* 47A:49, 1927.

9. G. Lemaître, "Spherical Condensations in the Expanding Universe," *Comptes Rendus de l'Academie des Sciences,* Paris, 196:903, 1933.

10. P. J. E. Peebles, "The Black-body Radiation Content of the Universe and the Formation of Galaxies," *Astrophysical Journal,* 142:1317, 1965.

11. Changbom Park, "Large N-body Simulations of a Universe Dominated by Cold Dark Matter," *Monthly Notices of the Royal Astronomical Society,* 242:59P, 1990; also unpublished work by Edmund Bertschinger and James Gelb, both at M.I.T.

12. J. Silk, "Fluctuations in the Primordial Fireball," *Nature,* 215:115, 1967; also, J. Silk, "Cosmic Blackbody Radiation and Galaxy Formation," *Astrophysical Journal,* 151:459, 1968.

Chapter Eight

1. A. G. Doroshkevich, Y. B. Zel'dovich, and I. D. Novikov, "The Origin of Galaxies in an Expanding Universe," *Soviet Astronomy,* 11:233, 1967.

2. J. P. Ostriker and L. L. Cowie, "Galaxy Formation in an Intergalactic Medium Dominated by Explosions," *Astrophysical Journal Letters,* 243:L127, 1981.

3. John Horgan, "Universal Truths," *Scientific American,* October 1990, p. 114.

4. R. Cowen, "Quasars: The Brightest and the Farthest," *Science News,* 139:276, May 4, 1991.

5. M. Mitchell Waldrop, "Cosmologists Begin to Fill in the Blanks," *Science,* 251:31, January 4, 1991.

6. Overbye, p. 352.

7. See A. Vilenkin, "Creation of Universes from Nothing," *Physics Letters B,* 117:25, 1982.

8. N. Turok, "Global Texture as the Origin of Cosmic Structure," *Physics Review Letters,* 63:2625, December 11, 1989.

9. For an overview on gravitational lensing, see Edwin L. Turner's excellent "Gravitational Lenses" in *Scientific American,* July 1988.

10. Horgan, Ibid., p. 115.

Chapter Nine

1. Columbia University sociologist Robert K. Merton explored the history of Newton's statement in *On the Shoulders of Giants: A Shandyan Postscript,* New York: Free Press, 1965.

2. Thomas S. Kuhn, 1977, pp. xii–xiii.

3. Thomas S. Kuhn, 1962, p. 110.

4. Regis, p. 224.

5. Scott Tremaine, "A Historical Perspective on Dark Matter," in *Dark Matter in the Universe,* Kormendy and Knapp, eds., Dordrecht, Holland: Reidel, 1987, pp. 547–549.

6. P. J. E. Peebles and Joseph Silk, "A Cosmic Book of Phenomena," *Nature,* 346:233–239, July 19, 1990.

Chapter Ten

1. See Stephen Myers and Emilio Picasso, "The LEP Collider," *Scientific American,* July 1990.

2. Michael D. Lemovich, "The Ultimate Quest," *Time,* April 16, 1990, p. 50.

3. For a good description of accelerator technology for nontechnical readers, see Leon Lederman and David N. Schramm, 1989.

4. Opal Collaboration, "A Combined Analysis of the Hadronic and Leptonic Decays of the Z^0," CERN EP/90–27, February 23, 1990.

5. John Boslough, "Worlds within the Atom," *National Geographic,* May 1985, p. 653.

6. Bertrand Russell, 1966, p. 45.

7. See Max Jammer, 1954.

8. See Bernard Jaffe, 1960.

9. Joseph Larmor, *Aether and Matter, Including a Discussion of the Influence of the Earth's Motion and Optical Phenomena,* Cambridge: Cambridge University Press, 1900.

Chapter Eleven

1. The pathfinding work was Bohr's "On the Constitution of Atoms and Molecules," *The London, Edinburgh, and Dublin Philosophical Magazine and Journal,* 26:1–25, July 1913. See also Bohr, 1960.
2. Bill Becher, "Pioneer of the Atom," *The New York Times Magazine,* October 20, 1957, p. 52.
3. See Werner Heisenberg, 1962, 1971.
4. See E. Schrödinger in *Quantum Theory and Measurement,* J. A. Wheeler and W. Zurek, eds., Princeton: Princeton University Press, 1983.
5. A good account of the development of quantum mechanics is given in Max Jammer, 1966.
6. See Helge S. Kragh, 1990, for a good description of Dirac and his science.
7. An early statement of the matter-antimatter problem is A. D. Sakharov's "Violation of CP Invariance, C Asymmetry and Baryon Asymmetry of the Universe," Jeremy Bernstein and Gerald Feinberg, eds., 1986; a later analysis is Steven Weinberg's "Cosmological Production of Baryons," *Physical Review Letters,* 42:850, 1978.
8. A good, not-too-technical explanation of quantum mechanics is in J. C. Polkinghorne, 1989. Polkinghorne, who was professor of Mathematical Physics at Cambridge University, resigned in 1979 to train for the ministry of the Church of England and is now an ordained priest.
9. In a remarkable paper in 1949, Fermi and C. N. Yang of the Institute for Advanced Studies had first hinted of the developing crisis in particle physics; see E. Fermi and C. N. Yang, "Are Mesons Elementary Particles?" in *The Physical Review,* December 15, 1949.
10. A representative sample of Gell-Mann's prodigious works during the early 1960s would include his "Form Factors and Vector Mesons" (with F. Zachariasen) in *The Physical Review,* 124:953, 1961; "Gauge Theories of Vector Particles" (with Sheldon L. Glashow) in *Annals of Physics,* 15:437, 1961; and "Symmetries of Baryons and Mesons," in *The Physical Review,* 125:1067, 1962.
11. Murray Gell-Mann, *The Eightfold Way: A Theory of Strong Interaction Symmetry,* Pasadena: California Institute of Technology, 1961.
12. Zweig's CERN report, 8419/TH.412, entitled "An SU3 Model for Strong Interaction Symmetry and its Breaking," was finally published in 1980 in *Developments in the Quark Theory of Hadrons: A Reprint Collection,* D. B. Lichtenberg and S. P. Rosen, eds., Nonantum, Mass.: Hadronic Press, pp. 22–101.
13. M. Gell-Mann, "A Schematic Model of Baryons and Mesons," in *Physics Letters,* 8:214, 1964.

Chapter Twelve

1. S. L. Glashow, "Partial Symmetries of Weak Interactions," *Physical Nuclear Physics (Intl.),* 22:579–88, February 1961.
2. P. W. Higgs, "Broken Symmetries, Massless Particles, and Gauge Fields," *Phys. Letters (Netherlands),* 12:132–3, September 15, 1964. See also by same

author: "Broken Symmetries and the Masses of Gauge Bosons," _Physical Review Letters,_ 13:508–9, October 19, 1964.

3. Charles C. Mann, "Armies of Physicists Struggle to Discover Proof of a Scot's Brainchild," _Smithsonian,_ March 1989, p. 110.

4. P. J. E. Peebles, "Primordial Helium Abundance and the Primordial Fireball II," in _Astrophysical Journal,_ 146:542, 1966; for an earlier analysis of the formation of the light elements, see Ralph A. Alpher, James W. Follin, Jr., and Robert C. Herman, "Physical Conditions in the Initial Stages of the Expanding Universe," _Physical Review,_ 92:1347, 1953.

5. _The Cosmos: Voyage Through the Universe,_ Alexandria, Va.: Time-Life Books, 1989, p. 98.

6. Leon M. Lederman and David Schramm, 1989, p. 160.

7. J. Ellis, K. Olive, D. V. Nanopoulos, and M. Srednicki, "Inflation Cries Out for Supersymmetry," _Physics Letters,_ 143B:429, 1984.

8. Penrose, 1989, p. 350.

Chapter Thirteen

1. See Ludovico Geymonat, _Galileo Galilei: A Biography and Inquiry into His Philosophy of Science,_ 1965. This book could as easily have been subtitled, "the making of the mind of the first physicist, a new kind of rationalist." See also William Wallace, _Galileo and His Sources,_ Princeton: Princeton University Press, 1984.

2. J. D. North, 1967, contains a good description of the mathematics used by Newton in the _Principia._

3. A. Einstein, "The General Theory of Relativity," _Sitzungsberichte der Preussischen Akad. d. Wissenschaften,_ 778, 1915, translated; also, "Cosmological Considerations on the General Theory of Relativity," _Sitzungsberichte der Perussischen Akad. d. Wissenschaften,_ 142, 1917, translated.

4. John Archibald Wheeler, 1990, is a good description of gravity and general relativity for the nontechnical reader.

5. Roland V. Eötvös, _Contributions to the Law of Proportionality of Inertia and Gravity,_ translated by J. Achtzehnter, et al., Institute for Nuclear Theory, Department of Physics, University of Washington, U.S. Department of Energy paper, no date; originally published in Annalen der Physik, 68:11, 1022, written with D. Pekár and J. Fekete. For an excellent overview of Eötvös's contributions to physics, which won several important prizes and played a role in the formation of general relativity, see R. H. Dicke, "The Eötvös Experiment," _Scientific American,_ 205:84–94, December 1961.

6. F. D. Stacey, G. J. Tuck, S. C. Holding, A. R. Maher, and D. Morris, "Constraints on the Planetary Scale Value of the Newtonian Gravitational Constant from the Gravity Profile Within a Mine," _Physical Review,_ D23:1683, 1981; also F. D. Stacey and G. J. Tuck, "Geophysical Evidence for Non-Newtonian gravity," _Nature,_ 292:230, 1981.

7. E. Fischbach, D. Sudarsky, A. Szafer, C. Talmage, and S. Aronson, "The Fifth Force," _Physics Review Letters,_ 56:3, 1986.

8. P. E. Boynton, D. Crosby, P. Ekstrom, A. Szumilo, "Search for an Intermediate-Range Composition-Dependent Force," *Physics Review Letters,* 58:1385, 1987.

9. For another positive fifth-force result, see D. H. Eckhardt, C. J Jekeli, A. R. Lazarewicz, A. J. Romaides, and R. W. Sands, "Tower Gravity Experiment: Evidence for Non-Newtonian Gravity," *Physics Review Letters,* 60:25, 1988.

10. In "Gravity Anomalies at the Nevada Test Site," an unpublished abstract, Thomas and colleagues Petr Vogel and Paul Kasameyer reported they had found an approximately 2.5 percent discrepancy from the normal gravity gradient of the Earth as predicted by a standard Newtonian model.

11. J. B. S. Haldane, "On Being the Right Size," in *Possible Worlds and Other Essays,* London: Chatto and Windus, 1927.

12. Pais, 1982, p. 179.

13. In 1971 four extremely accurate cesium clocks were sent aboard commercial jetliners on around-the-world journeys in the opposite direction. Because of the eastward spin of the earth and its atmosphere, the plane traveling eastward would go faster, and its clock was expected to lose 315 nanoseconds (billionths of a second). It lost 332 nanoseconds, well within experimental limits. Other atomic clocks sent into orbit have repeatedly recorded similar differences.

14. John Boslough, "The Enigma of Time," *National Geographic,* 177:3, March 1990.

15. For a good history of the development of Western ideas about the concept of time, see Stephen Toulmin and June Goodfield, 1965.

16. Edward T. Hall, 1984.

17. J. T. Fraser, 1987, p. 67.

Chapter Fourteen

1. Stephen Hawking, "The Direction of Time," *New Scientist,* July 9, 1987, p. 46.

2. Ibid.

3. John Boslough, 1985, pp. 19–31.

4. S. W. Hawking and R. Penrose, "The Singularities of Gravitational Collapse and Cosmology," *Proceedings of the Royal Society of London,* A314:529, 1969.

5. S. W. Hawking, "Black Hole Explosions?" *Nature,* 248:30, 1974. For a nontechnical account, see S. W. Hawking, "The Quantum Mechanics of Black Holes," *Scientific American,* January 1977, pp. 34–40.

6. Stephen W. Hawking, 1988, p. 151.

7. See S. W. Hawking, "The Boundary Conditions of the Universe," *Pontificae Academiae Scientarium Scripta Varia,* 48:563, 1982; and "The Quantum State of the Universe," *Nuclear Physics B,* 239:257, 1984; also J. B. Hartle and S. W. Hawking, "Wave Function for the Universe," *Physical Review D,* 28:2960, 1983.

8. Hawking, 1988, p. 140.

9. Hawking, *Is the End in Sight for Theoretical Physics? An Inaugural Lecture,* Cambridge: Cambridge University Press, 1980, pp. 25–26.

Chapter Fifteen

1. Roger Penrose, "Gravitational Collapse and Space-Time Singularities," *Physical Review Letters,* 14:57, 1965.

2. S. W. Hawking and R. Penrose, "The Singularities of Gravitational Collapse and Cosmology," *Proceedings of the Royal Society of London,* A314:529, 1969.

3. Penrose's earliest published work on twistor theory was "Twistor Quantization and the Curvature of Spacetime," in the *International Journal of Theoretical Physics,* 1:61, 1968.

4. Roger Penrose, 1989.

5. S. W. Hawking and M. Rocek, eds., *Superspace and Supergravity: Proceedings of the Nuffield Workshop,* Cambridge: Cambridge University Press, 1981.

6. Y. Nambu, Unpublished paper, *Copenhagen Symposium,* 1970.

7. M. B. Green and J. H. Schwarz, *Physics Letters,* 149B:117, 1984.

8. For a thorough description of the concept for the not-too-technical reader, see Michael B. Green's "Superstrings," in *Scientific American,* September 1986.

9. J. I. Merritt, "Toward a Theory of Everything," *Princeton Alumni Weekly,* April 9, 1986, p. 26.

10. Paul Ginsparg and Sheldon Glashow, "Desperately Seeking Superstrings," *Physics Today,* May 1986, p. 7.

11. E. P. Tryon, "Is the Universe a Vacuum Fluctuation?" *Nature,* 246:396, 1973.

12. Nigel Calder, 1979, pp. 127–131.

13. S. Coleman, "Why There Is Nothing Rather Than Something," *Nuclear Physics* B310:643–668, 1988. See also: Bertram Schwarzschild, "Why Is the Cosmological Constant So Very Small," *Physics Today,* March 1989, pp. 21–24; nontechnical readers should see John Horgan, "Measuring Eternity," *Scientific American,* December 1990, pp. 16–17; and Edmund Bertschinger, "Einstein's Blunder Resurrected," *Nature,* December 20–27, 1990, pp. 675–676.

14. David H. Freedman, "Maker of Worlds," *Discover,* July 1990, p. 52.

15. M. S. Morris and K. S. Thorne, "Wormholes in Spacetime and Their Use for Interstellar Travel: A Tool for Teaching General Relativity," *American Journal of Physics,* 56:395–412, May 1988.

16. Michael S. Morris, Kip S. Thorne and U. Yurtsever, "Wormholes, Time Machines, and the Weak Energy Condition," *Physical Review Letters,* 61:1446, September 26, 1988.

Chapter Sixteen

1. This point was first made by Dennis Overbye in *Lonely Hearts of the Cosmos,* 1991.

2. Dennis Overbye, "Messenger at the Gates of Time," *Science 81,* June 1981, pp. 61–62.

3. John Boslough, "Inside the Mind of John Wheeler," *Reader's Digest,* September 1986.

4. See J. Barrow and F. Tipler, 1986, for a discussion of this fascinating sidelight to cosmology.

5. Heinz R. Pagels, "A Cozy Cosmology," *The Sciences,* 25:2, March/April 1985.

6. Wheeler amplified his observer-participator ideas in "Information, Physics, Quantum: The Search for Links," Princeton University and University of Texas preprint, February 1990.

7. Bill Becher, "Pioneer of the Atom," *The New York Times Magazine,* October 20, 1957, p. 57.

8. Italo Calvino, 1985.

9. For a readable discussion of Alfvén's and other alternative cosmologies, see Anthony L. Peratt's "Not with a Bang" in *The Sciences,* January/February 1990.

10. H. C. Arp, G. Burbridge, F. Hoyle, J. V. Narlikar, and N. C. Wickramasinghe, "The Extragalactic Universe: An Alternative View," *Nature,* 346:807, August 30, 1990.

11. Halton Arp, "Letters," *Science News,* July 27, 1991, p. 51.

12. Ibid.

13. See John Maddox's "Down with the Big Bang," in *Nature,* 340:425, August 10, 1989. In Robert Oldershaw's "What's Wrong with the New Physics?" in the British publication, *New Scientist,* December 22–29, 1990, the author discusses the tendency of modern cosmologists to develop hypotheses that cannot be tested and to ignore experimental data that contradict their ideas.

14. Isaac Newton, *Principia,* Cambridge: Cambridge University Press, 1687.

15. For more on Gödel, see John Dawson, "Kurt Gödel in Sharper Focus," *The Mathematical Intelligencer,* 6:9, 1984. Gödel had a number of novel and interesting ideas. In his "An Example of a New Type of Cosmological Solutions of Einstein's Field Equations of Gravitation" in *Reviews of Modern Physics,* 21:447, 1947, he presents a theory of time travel.

Books are not absolutely dead things.
—JOHN MILTON

Only books and articles likely to be of interest to the general reader are listed here. See the Notes for technical books and scientific papers.

Aristotle. *The Complete Works of Aristotle.* Jonathan Barnes, ed. Princeton: Princeton University Press, 1984.

Armitage, Angus. *Copernicus: The Founder of Modern Astronomy.* New York: Barnes, 1962.

Barrow, John D., and Frank J. Tipler. *The Anthropic Cosmological Principle.* Oxford: Oxford University Press, 1988.

Bernstein, Jeremy. *Einstein.* New York: Viking, 1973.

Bernstein, Jeremy, and Gerald Feinberg, eds. *Cosmological Constants: Papers in Modern Cosmology.* New York: Columbia University Press, 1988. From Einstein to cosmic strings.

Blum, Harold F. *Time's Arrow and Evolution.* Princeton: Princeton University Press, 1968. Humanity and thermodynamics.

Bohr, Niels. *Atomic Physics and Human Knowledge.* New York: Wiley, 1960.

Bondi, Hermann. *Assumption and Myth in Physical Theory.* Cambridge: Cambridge University Press, 1967.

Boorstin, Daniel. *The Discoverers.* New York: Random House, 1983. Fine opening section on the discovery of time.

Boslough, John. "The Enigma of Time." *National Geographic,* 177:3, March 1990.

———. "Searching for the Secrets of Gravity." *National Geographic,* 175:5, May 1989.

———. *Stephen Hawking's Universe.* New York: William Morrow, 1985.

Bronowski, Jacob. *The Ascent of Man.* Boston: Little, Brown, 1973. A classic work of humanity's self-discovery.

Butterfield, Herbert. *The Origins of Modern Science,* 1300–1800. New York: Macmillan, 1951. A classic study.

Calder, Nigel. *Einstein's Universe.* New York: Greenwich House, 1979. A crystal-clear exposition of relativity and its offshoots.

Calvino, Italo. *Mister Palomar.* New York: Harcourt, Brace & Javonovich, 1985. How *should* one view the universe?

Carrigan, Richard A., Jr., and W. Peter Trower, eds. *Particle Physics in the Cosmos.* New York: Freeman, 1989.

Clark, Ronald W. *Einstein: The Life and Times.* New York: World Publishing, 1971. A classic biography for the nontechnical reader.

Cohen, Nathan. *Gravity's Lens.* New York: Wiley, 1988.

247

Cornell, James, ed. *Bubbles, Voids, and Bumps in Time: The New Cosmology.* Cambridge: Cambridge University Press, 1988. A collection of popular articles by modern cosmologists.

Davies, Paul. "Cosmogenesis." *Creation,* May/June 1990. Suggests that we inhabit a progressive rather than degenerative universe as described by the second law of thermodynamics.

———. *God and the New Physics.* New York: Simon & Schuster, 1983.

———. *Superforce.* New York: Simon & Schuster, 1984. What really caused the big bang?

Davies, P. C. W., and J. Brown, eds. *Superstrings: A Theory of Everything?* Cambridge: Cambridge University Press, 1988.

Dawson, John, trans. and ed. "Discussion on the Foundation of Mathematics." *History and Philosophy of Logic 5,* 1984. A reasonably accessible introduction to Kurt Gödel's incompleteness theorem.

Dirac, P. A. M. "The Evolution of the Physicist's Picture of Nature." *Scientific American,* May 1963. The evolution of Dirac's picture of nature.

Dukas, Helen, and Banesh Hoffman, eds. *Albert Einstein: The Human Side.* Princeton: Princeton University Press, 1979. Words of wisdom on a number of subjects from the master of the universe in both German and English.

Dyson, Freeman. *Infinite in All Directions.* New York: Harper & Row, 1985. Lecture series by the scientist-humanist.

Eddington, A. S. *The Expanding Universe.* New York: Macmillan, 1933. How the universe was perceived five years after Hubble's discovery of galactic recession.

———. *The Nature of the Physical World.* New York: Macmillan, 1928. How the universe was perceived on the verge of Hubble's discovery of galactic recession.

Einstein, Albert. *The Evolution of Physics.* With Leopold Infeld. New York: Simon & Schuster, 1938.

———. *The Meaning of Relativity.* Princeton: Princeton University Press, 1955. One of Einstein's most accessible works on the subject, based on lectures of 1922.

Feinberg, Gerald. *What is the World Made Of? Atoms, Leptons, Quarks and Other Tantalizing Particles.* New York: Doubleday, 1977.

Ferris, Timothy. *Coming of Age in the Milky Way.* New York: William Morrow, 1988. A history of humanity's struggle to understand the universe.

Ferris, Timothy, ed. *The World Treasury of Physics, Astronomy and Mathematics.* Boston: Little, Brown, 1991. Wide-ranging readings.

Feynman, Richard P. *The Feynman Lectures on Physics.* 3 Volumes. Reading, Massachusetts: Addison-Wesley, 1963.

———. *QED: The Strange Theory of Light and Matter.* Princeton: Princeton University Press, 1983. A good, basic introduction to quantum electrodynamics.

———. *Surely You're Joking, Mr. Feynman!* New York: Norton, 1985. Delightful reading from one of our greatest physicists.

Fraser, Julius T. *Time as Conflict.* Basel: Birkhäuser, 1978. An authoritative, often compelling, often frustrating work from a leading philosopher of time.

————. *Time: The Familiar Stranger.* Boston: University of Massachusetts Press, 1987.

French, A. P. *Newtonian Mechanics.* New York: Norton, 1971.

Gamow, George. *One, Two, Three . . . Infinity.* New York: Viking, 1947. Inspired thousands of schoolchildren in the early days of the big bang.

————. *Mr. Tompkins in Wonderland.* New York: Macmillan Co., 1946.

Gell-Mann, M., and E. P. Rosenbaum. "Elementary Particles." *Scientific American,* July 1957. An authoritative introduction to the subject.

Gell-Mann, Murray, and Yuval Ne'eman. *The Eightfold Way.* New York: Benjamin, 1964.

Geymonat, Ludovico. *Galileo Galilei: A Biography and Inquiry into His Philosophy of Science,* trans. Stillman Drake. New York: McGraw-Hill, 1965.

Ginsparg, Paul, and Sheldon Glashow. "Desperately Seeking Superstrings." *Physics Today,* May 1966. Anybody who reads this will see why the authors don't like the latest theory of everything.

Glashow, Sheldon. "Quarks with Color and Flavor." *Scientific American,* April 1974. A basic introduction.

Green, Michael B. "Superstrings." *Scientific American,* September 1986. A good introduction to the latest theory of everything.

Gribbin, John. *In Search of the Big Bang.* New York: Bantam, 1986.

Gulkis, Samuel, Philip M. Lubin, Stephan S. Meyer, and Robert F. Silverberg. "The Cosmic Background Explorer." *Scientific American,* January 1990. A good general description of the instruments on board.

Guth, Alan H. "The Birth of the Cosmos" in D. E. Osterbroch and P. H. Raven, eds., *Origins and Extinctions.* New Haven: Yale University Press, 1988. This article is being expanded into a book.

Guth, Alan H., and Paul J. Steinhardt. "The Inflationary Universe." *Scientific American,* 250, May 1984. Probably the most accessible article on the subject.

Hall, Edward T. *The Dance of Life: The Other Dimension of Time.* Garden City: Anchor/Doubleday, 1984. How concepts of time vary from culture to culture.

Hawking, Stephen W. *A Brief History of Time.* New York: Bantam, 1988.

————. "The Direction of Time." *New Scientist,* July 9, 1987. Hawking's own account of his flip-flop on the direction of time.

————. "The Quantum Mechanics of Black Holes." *Scientific American,* January 1977. A generally accessible account for the general reader.

Hawking, S. W., and W. Israel, eds. *General Relativity: An Einstein Centenary Survey.* Cambridge: Cambridge University Press, 1979. Readings from a number of physicists.

Heisenberg, Werner. *Encounters with Einstein.* Princeton: Princeton University Press, 1989. An appraisal in nine essays of the scientific method in the twentieth century.

————. *Physics and Beyond.* New York: Harper & Row, 1971.

————. *Physics and Philosophy.* New York: Harper, 1962.

Hoffmann, Banesh, and Helen Dukas. *Albert Einstein: Creator and Rebel.* New York: Viking, 1972.

Hofstadter, Douglas R. *Gödel, Escher, Bach: An Eternal Golden Braid.* New York: Vantage, 1980. Profoundly interesting.

Hoyle, Fred. *Galaxies, Nuclei, and Quasars.* New York: Harper, 1965.

———. *The Nature of the Universe.* New York: Harper, 1950. The universe without the big bang.

———. *Steady-state Cosmology Re-visited.* Cardiff: University College Cardiff Press, 1980.

Hubble, E. P. *The Observational Approach to Cosmology.* Oxford: Oxford University Press, 1937.

———. *The Realm of the Nebulae.* New Haven: Yale University Press, 1985.

Jaffe, Bernard. *Michelson and the Speed of Light.* New York: Doubleday, 1960.

Jammer, Max. *Concepts of Space: The History of Theories of Space in Physics.* Cambridge, Massachusetts: Harvard University Press, 1954.

———. *The Conceptual Development of Quantum Mechanics.* New York: McGraw-Hill, 1966. A good account of the theory's development.

Koyré, Alexandre. *Metaphysics and Measurement: Essays in Scientific Revolution.* Cambridge, Massachusetts: Harvard University Press, 1968.

Kragh, Helge S. *Dirac: A Scientific Biography.* Cambridge: Cambridge University Press, 1990. A fine account of the man and his work.

Krauss, Lawrence M. "Dark Matter in the Universe." *Scientific American,* December 1986. A condensed explanation.

———. *The Fifth Essence: The Search for Dark Matter in the Universe.* New York: Basic Books, 1989.

Kuhn, Thomas S. *The Copernican Revolution.* Cambridge, Massachusetts: Harvard University Press, 1970.

———. *The Essential Tension.* Chicago: University of Chicago Press, 1977. Kuhn explains how he arrived at scientific revolutions.

———. *The Structure of Scientific Revolutions.* Chicago: University of Chicago Press, 1970. Second edition of the book that redefined the history of science.

Landes, David S. *Revolution in Time: Clocks and the Making of the Modern World.* Cambridge, Massachusetts: Harvard University Press, 1983.

Lederman, Leon M., and David N. Schramm. *From Quarks to the Cosmos.* New York: Scientific American Library, 1989. An invaluable resource for understanding the relationship between the very large and the very small.

Leibniz, Gottfried Wilhelm. *Philosophical Writings,* G. H. R. Parkinson, ed., and trans. Mary Morris and G. H. R. Parkinson. London: Dent, 1973.

Lemaître, G. *The Primeval Atom.* New York: D. Van Nostrand, 1950. One of the earliest expositions of the big bang.

Lerner, Eric J. *The Big Bang Never Happened: A Startling Refutation of the Dominant Theory of the Origin of the Universe.* New York: Times Books, 1991. Promulgates an alternative theory that the universe has been shaped by electric currents operating in a plasma.

Leslie, John, ed. *Physical Cosmology and Philosophy.* New York: Macmillan, 1990. Engaging musings from George Gamow, Stephen Jay Gould and others.

Lightman, Alan, and Roberta Brawer, eds. _Origins: The Lives and Worlds of Modern Cosmologists._ Cambridge, Massachusetts: Harvard University Press, 1990. Interviews with the leading lights of cosmology.

Maddox, John. "Down with the Big Bang." _Nature,_ August 10, 1989.

Misner, Charles W. "Cosmology and Theology." In _Cosmology, History and Theology,_ Wolfgang Yourgrau and Allen D. Breck, eds. New York: Plenum Press, 1977. A good explanation of the horizon problem.

Morris, Richard. _The Edges of Science: Crossing the Boundary from Physics to Metaphysics._ New York: Prentice Hall, 1990. A clear exposition of the problems faced by modern physics.

————. _Time's Arrows._ New York: Simon & Schuster, 1985. Examines the problem of time from a physics and human perspective.

Munitz, M. K., ed. _Theories of the Universe._ New York: Free Press, 1957. Papers in cosmology back to Aristotle.

Myers, Stephen, and Emilio Picasso. "The LEP Collider." _Scientific American,_ July 1990.

Neumann, John von. _The Computer and the Brain._ New Haven: Yale University Press, 1958.

North, J. D. _The Measure of the Universe._ Oxford: Clarendon Press, 1965. A good history of cosmology.

————. _Isaac Newton._ Oxford: Oxford University Press, 1967. A good account of Newton's mathematics for the general reader.

Oldershaw, Robert. "What's Wrong with the New Physics?" _New Scientist,_ December 22–29, 1990. Nontechnical observations about the ways theoretical physicists flout the so-called scientific method by ignoring data that contradict their theories and devising theories that cannot be tested.

Overbye, Dennis. _Lonely Hearts of the Cosmos._ New York: HarperCollins, 1991. Popularly written and entertaining account of cosmology, focusing on Allan Sandage.

Pagels, Heinz. _The Cosmic Code._ New York: Simon & Schuster, 1982. A clear, often quirky, always fascinating look at the quantum world.

————. _Perfect Symmetry._ New York: Simon & Schuster, 1985.

Pais, Abraham. _Subtle is the Lord: The Science and the Life of Albert Einstein._ London: Oxford University Press, 1982.

Peat, F. David. _Superstrings and the Search for the Theory of Everything._ Chicago: Contemporary Books, 1988.

Peebles, James. _The Large-Scale Structure of the Universe._ Princeton, N.J.: Princeton University Press, 1980.

————. _Physical Cosmology._ Princeton, N.J.: Princeton University Press, 1971.

Penrose, Roger. _The Emperor's New Mind: Concerning Computers, Minds, and the Laws of Physics._ Oxford: Oxford University Press, 1989. A hard but rewarding look at quantum physics, theories of everything and artificial intelligence.

Peratt, Anthony L. "Not with a Bang: The Universe May Have Evolved from a Vast Sea of Plasma." _The Sciences,_ January/February 1990. An excellent nontechnical article questioning the big bang and detailing the plasma universe of Swedish Nobel laureate Hannes Alfvén.

Peterson, Ivars. "State of the Universe: If Not with a Big Bang, Then What?" *Science News,* April 13, 1991. A good summary of opposing views.

Polkinghorne, J. C. *The Quantum World.* Princeton: Princeton University Press, 1985.

Popper, Karl. *The Logic of Scientific Discovery.* New York: Harper & Row, 1968. A classic study from the renowned philosopher of science.

Regis, Ed. *Who Got Einstein's Office? Eccentricity and Genius at the Institute for Advanced Study.* Reading, Massachusetts: Addison-Wesley, 1987.

Rubin, Vera C. "Dark Matter in Spiral Galaxies." *Scientific American,* December 1988. A good general discussion of its discovery and implications.

Russell, Bertrand. *The Wisdom of the West.* London: Rathbone Books, 1959.

Rutherford, Ernest. *The Collected Papers of Lord Rutherford of Nelson,* James Chadwick, ed. 3 Volumes. New York: Interscience, 1962.

Sambursky, S. *The Physical World of the Greeks.* Translated from the Hebrew by Merton Dagut. New York: Collier, 1962.

Sandage, Allan. *The Hubble Atlas of Galaxies.* Washington, D.C.: Carnegie Institution, 1961.

Sarton, George. *A History of Science.* New York: Norton, 1970.

Schwarz, John H. "Completing Einstein." *Science 85,* November 1985. A basic explanation of superstrings by one of the theory's creators.

Silk, Joseph. *The Big Bang.* San Francisco: Freeman, 1989. Discusses the big issues in cosmology, such as dark matter.

Spitzer, Lyman. *Searching Between the Stars.* New Haven: Yale University Press, 1982. The father of nuclear fusion muses about interstellar matter.

Taubes, Gary. *Nobel Dreams.* New York: Random House, 1986. Lively account of Carlo Rubbia's steamroller approach to subatomic physics.

Toulmin, Stephen, and June Goodfield. *The Architecture of Matter.* New York: Harper & Row, 1962.

———. *The Discovery of Time.* New York: Harper & Row, 1965. Volume Three in a classic four-volume series on the ancestry of science.

———. *The Fabric of the Heavens.* New York: Harper & Row, 1961.

Trefil, James. *Reading the Mind of God: In Search of the Principle of Universality.* New York: Anchor/Doubleday, 1989. Why we need a principle of universality.

Tucker, Wallace, and Karen Tucker. *The Dark Matter: Contemporary Science's Quest for the Mass Hidden in Our Universe.* New York: William Morrow, 1988.

Turner, Edwin L. "Gravitational Lenses." *Scientific American,* July 1988. An excellent nontechnical overview.

Wallace, Alfred Russel. *Man's Place in the Universe.* New York: McClure Phillips Co., 1903.

Wallace, William. *Galileo and His Sources.* Princeton, N.J.: Princeton University Press, 1984.

Weinberg, Steven. *The First Three Minutes.* New York: Bantam, 1984. The classic introduction to the early universe.

Weisskopf, Victor W. *Knowledge and Wonder.* Cambridge, Massachusetts: M.I.T. Press, 1979.

Westergren, Göran. *Time: Experiences, Perspectives, and Coping Strategies.* Stockholm: Almqvist and Wiksell Intl., 1990.

Wheeler, John Archibald. *Albert Einstein, 1879–1955: A Biographical Memoir.* Washington, D.C.: National Academy of Sciences, 1980.

———. *A Journey into Gravity and Spacetime.* New York: Scientific American Library, 1990. A wonderful excursion with an engaging teacher.

White, Andrew Dickson. *A History of the Warfare of Science with Theology in Christendom.* New York: Macmillan, 1965. The battle rages on, with science holding the upper hand.

Whitehead, Alfred North. *The Concept of Nature.* London: Cambridge University Press, 1964.